CONSENSUS FORMATION IN HEALTHCARE ETHICS

Philosophy and Medicine

VOLUME 58

EUROPEAN STUDIES IN PHILOSOPHY OF MEDICINE 2

The titles published in this series are listed at the end of this volume.

CONSENSUS FORMATION IN HEALTHCARE ETHICS

Edited by

HENK A.M.J. TEN HAVE

Catholic University of Nijmegen,
School of Medical Sciences,
Nijmegen, The Netherlands

and

HANS-MARTIN SASS

Ruhr Universität, Institut für Philosophie, Bochum, Germany;
Kennedy Institute of Ethics, Georgetown University,
Washington, D.C., U.S.A.

KLUWER ACADEMIC PUBLISHERS
DORDRECHT / BOSTON / LONDON

A C.I.P Catalogue record for this book is available from the Library of Congress.

ISBN 978-90-481-4993-3

Published by Kluwer Academic Publishers,
P.O. Box 17, 3300 AA Dordrecht, The Netherlands

Sold and distributed in North, Central and South America
by Kluwer Academic Publishers,
P.O. Box 358, Accord Station, Hingham, MA 02018-0358, U.S.A.

In all other countries, sold and distributed
by Kluwer Academic Publishers, Distribution Center,
P.O. Box 322, 3300 AH Dordrecht, The Netherlands

Printed on acid-free paper

TABLE OF CONTENTS

ACKNOWLEDGEMENTS

Most of the essays in this volume were first presented in 1990 at the Fourth Annual Meeting of the European Society for Philosophy of Medicine and Healthcare (ESPMH), in Maastricht, The Netherlands. That conference was originally scheduled to convene at the Universität in Bochum, Germany. Due to concerted actions by a network of so-called "alternative groups", who intended to interrupt the meeting and to harass participants, productive and peaceable deliberations could not be guaranteed in Bochum. The meeting was, within the space of a few hours, rescheduled to convene in Maastricht. We take this opportunity to thank the organizers in Bochum and Maastricht for making a successful and peaceable conference possible. We also wish to thank the speakers, all of whom have revised their manuscripts for publication, as well as those who were invited to contribute to this volume after the conference adjourned. Our appreciation is also extended to the co-editors of the *Philosophy and Medicine* series, Stuart F. Spicker and H. Tristram Engelhardt, Jr., for working with us on this volume, to an anonymous reviewer, and to Ms. Marian Poulissen for editing the text and preparing the Index.

Henk A.M.J. ten Have
Hans-Martin Sass

HENK A.M.J. TEN HAVE
HANS-MARTIN SASS

INTRODUCTION: CONSENSUS FORMATION IN HEALTHCARE ETHICS

CONSENSUS AND BIOETHICS

In philosophy, consensus formation is not an immediate concern. Dissensus, disagreement and heterogeneity seem to dominate along with fundamentally incompatible systems and schools of thought. How can a discipline in search of wisdom and truth produce multiple, divergent and mutually exclusive theoretical systems? How can a discipline that invented logical analysis and argumentative methodology create an increasing number of inconsistent views? Students participating in international exchange programs learn that a philosophical education at the universities in various countries requires reading different books, studying different periods in history, and making themselves familiar with the works of different philosophers. Nonetheless, there is at the moment a greater exchange of ideas and more books than ever before in the history of philosophy. French philosophers are teaching in North-American universities; German philosophers are involved in British research programs; and an increasing number of European students spend some time pursuing educational programs in "foreign" universities. At the same time, analytical philosophy, structuralism, and critical theory have not been succeeded by any one dominating European school of thought. But the present state of fragmentation, eclecticism, and 'bricolage', has certainly not led to a decline of interest in philosophy. Although the philosophical "answers" may differ, the perplexing questions have remained the same. Because of this common set of questions, philosophers continue to try to formulate general answers, identify primary principles, and design common frameworks.

This paradoxical situation is even more obvious in bioethics; many schools and approaches flourish in practice and are reflected in the literature: applied ethics, hermeneutical ethics, casuistry, clinical ethics, narrative ethics, feminist ethics. This diversity of approaches and methods does not, however, discredit bioethics, but serves as a stimulus to involve

H.A.M.J. ten Have and H.-M. Sass (eds.), Consensus Formation in Healthcare Ethics, 1–14.

more and more participants in the debates over the moral dimensions of healthcare. The more heterogeneity, the greater the motive to attempt to locate some ground for common understanding. This is also the critical function of philosophy: it should never take for granted any existing consensus; it should always persist in asking troubling questions about generally shared assumptions and opinions. The same holds for bioethics; its function also is critically to examine and analyze the prevailing consensus regarding moral issues. It can be argued that as long as there is a general consensus about the moral perplexities in healthcare, the need for explicit reflection on the moral dimension of healthcare (and therefore for bioethical activities) should remain rather limited.

Bioethics as a discipline is flourishing because a moral consensus has evanesced and is itself questioned. Physicians, philosophers, lawyers, and theologians are engaged in bioethics, since they no longer concur with the moral consensus which prevailed in the past. But at the same time bioethicists also are instrumental in trying to develop a new moral consensus. They are consulted with the hope that they can help to create and formulate a consensus regarding the most complicated moral problems in healthcare. Approaches to consensus formation, developed in part over the last decade to establish standards of clinical practice, are now extended to moral issues. These approaches can be observed at various levels: institutional (with the development of protocols by healthcare ethics committees), professional (with the creation of ethics advisory committees and the appointment of officers for ethics standards, representing national medical associations), national (with the creation of national ethics committees in many countries), and even international. An interesting example of the latter is the effort of the Council of Europe to develop a *Convention on human rights and biomedicine*. In July 1994, the Council published the first *Draft Convention for the protection of human rights and dignity of the human being with regard to the application of biology and medicine: Bioethics Convention* [8]. The text has been prepared by the Steering Committee on Bioethics (CDBI: Comité Directeur pour la Bioéthique). After lengthy negotiations among the government delegates, a preliminary trans-national consensus statement was finally offered for public debate among the Council's 40 member states. Based on the numerous comments received regarding the first draft, a second draft was published in June, 1996 [6]. This draft was approved by the Council's Committee of Ministers and the Parliamentary Assembly later in the same year. All forty member states are now in the process of debating (and possibly ratifying) the Convention at the national level [13].

The Convention identified basic principles necessary for the application of medicine and the life sciences. It set out to protect the dignity and identity of all human beings: "The interests and welfare of the human being shall prevail over the sole interest of society or science" (art.2). It also requests that appropriate measures be taken to provide equitable access to healthcare of appropriate quality. The principle of respecting the free and informed consent of the person is clearly stressed. The issue of interventions on persons unable to provide an informed consent has been controversial for a long time; in the first draft, the exact formulation of the relevant article remained open. Now it states that interventions may be carried out on persons with impaired decision-making capacities, but only for their benefit, and only if minimal risks and minimal burdens are imposed. Consensus apparently exists over a broad range of issues. Privacy and free access to information are defined as rights, in article 10. Discrimination against a person on the basis of his or her genetic heritage is prohibited. Sex selection in medically assisted procreation is prohibited (except to avoid serious hereditary, sex-related diseases). Financial gain from using the human body and its parts is explicitly prohibited.

The more controversial issues in bioethics are not addressed in this Convention, however. An exception is research on embryos *in vitro*. Article 18 states, that "The creation of human embryos for research purposes is prohibited." Specific problems can be elaborated in special protocols that modify the Convention. The first draft announced that protocols are in preparation for organ transplantation and experimentation with human subjects, but the final version refers only to the possibility of formulating experimental protocols. Recently, a proposed draft protocol relating to research on the human embryo and fetus was published [4].

The European Convention on Human Rights and Biomedicine (the official title), is a well-intended and carefully prepared document that may stand as a landmark in the evolution of bioethics in Europe. It builds on the earlier foundations of the Universal Declaration of Human Rights as well as on the European Treaty for the Protection of Human Rights and Fundamental Freedoms. It identifies basic moral principles and moral procedures. For ethicists from North-western Europe the Convention could be disappointing due to the general character of its formulations. But given the status of bioethics throughout Europe, in particular the wide variety of theories and practices, the lack of bioethics education in many countries, the embryonic state of many procedures and committees in hospitals, as well as the poor quality of public debate, this Convention should lead to more concerted approaches to enhance the sophistication of bioethics in all European nations. At the same time, the above-mentioned problem remains: What is the importance and role of bioethics *vis-à-vis* this trans-national moral consensus?

THE NEED FOR CONSENSUS

In 1670, Baruch Spinoza's *Tractatus Theologico-Politicus* described the personal and societal benefits of not enforcing consensus on the specifics of religious dogma, such as the resurrection of Jesus, His miracles, or even His existence [12]. Spinoza argued that the individual's freedom to reason, to judge, and to assess traditional systems of orientation and belief are the foundation not only for individual freedom of conscience but also for a peaceable society. By contrast, consensus in matters of orientation and belief does not serve as such a foundation. With this model for strengthening individual conscience, Spinoza argued that modern societies would be stronger and more stable if based on individual responsibility and commitment rather than a consensus of ideas enforced by political or religious authorities.

Indeed, an imposed consensus turns out to be of no great benefit to individuals, groups, churches, and societies in resolving and avoiding conflicts, given the lessons of history regarding the religious and ideological wars in Europe and elsewhere. In healthcare we do not need to consent to specific religious beliefs, or even to pledge allegiance to schools of scientific inquiry and explanation in order to advocate good clinical practice: a commitment to help and to care is enough, no matter how such virtues, principles, and actions are rationally or emotionally supported.

But Spinoza's model did not work and does not work without a consensus regarding certain moral principles (such as *libertas*, *securitas*, and *solidaritas)* that hold the fabric of society together, no matter how they are supported individually by religious belief, or by humanist or utilitarian reasoning. What if consensus regarding those principles is not self-evident within a community? What if we face strong societal dissent in practical matters like security, solidarity, defense, or healthcare financing? Do we not at least have to agree that maintaining a peaceable society and protecting civil rights and responsibilities in the face of dissent among individuals, churches, and other groups in various fragmented societies are necessary goals to realize the various values held by individuals?

But even such goals presuppose consensus regarding certain basic values as well as certain mid-level principles, for which we have to seek consensus in order to support a peaceable community of freely contracting self-determined and self-responsible fellow humans. Respect for persons, human dignity, autonomy, and the responsibility to act in a spirit of solidarity with others seem to be necessary conditions to shape a universal consensus. Without them we are not able to resolve either simple or complex issues of mutual concern by negotiating with each other as "moral strangers." General consensus on the importance of principles like autonomy does not eliminate the need to negotiate the scope and limits of autonomy, or to achieve a precise definition of human dignity, e.g., whether, or to what extent unborn human life or brain-dead human life should be protected. Among mid-level principles essential to a consensus in healthcare matters are (say "the principlists") the principles of non-maleficence, beneficence, and respect for persons. If we do not have a consensus concerning the importance, perhaps even unavoidability of these principles, how can we begin the processes of forging mutual understanding and forming consensus on how to proceed further?

In addition, certain issues are crucially important for individual conscience, and yet they do not achieve societal consensus, at least in post-industrial society: contraception, abortion, organ donation, xenografts, criteria for death, and physician assistance in dying. Can we import the principle of *subsidiarity* (that the state should only interfere and make regulations when the issues cannot be resolved at a lower level of decisionmaking [11]) from social ethics in order to allow for responsible bioethical decisionmaking by those most immediately involved such as parents, family members, friends, and others who desire to provide secular or religious moral reasons? If we employ the principle of *subsidiarity* in moral concern and responsibility regarding contraception, can we do so also for issues like abortion, or only for certain requests for abortion? Should we leave it to those in need to accept xenografts or assistance in dying? When consensus on *content* cannot be achieved, then perhaps the focus has to shift towards shaping consensus on *processes* of toleration and towards negotiating *pragmatic* solutions.

The distinction between procedural and substantive consensus explains why consensus has become a central topic in bioethics and why it is fundamentally problematic. The importance of consensus is related, according to Caws [5], to changes in medical practice. In present-day medicine there is an increasing need of decision-making at the collective level. Now that healthcare professionals need to arrive at decisions as a team, they engage in collective deliberations intending to agree on a decision, perhaps not the best decision in the opinion of each deliberator, but at least the most acceptable to all. Jennings [10] argues that since morality is more and more understood as a socially embedded practice, the importance of consensus has increased. Moral agents need to construct consensus because there is no other basis available for the authority of moral claims. Moreover, emphasis is now on processes of deliberation through which agreement will emerge. Precisely this procedural interpretation, however, illustrates why consensus is fundamentally problematic. The fact that during deliberations the moral opinion of all deliberating persons converge on a particular conclusion has no bearing whatsoever on the *moral rightness* of that view. The social fact that moral agents agree does not in itself provide evidence of the rightness of the matter upon which agreement exists. What, then, is the moral weight or normative significance of consensus?

THE NEED FOR CONSENSUS IN HEALTHCARE

In postmodern societies, given the absence of authoritative rulers, priests, or prophets, morality is local, transitory, and fragmentary [1]. Given the diversity of moral visions, those who live within pluralistic societies have nonetheless a need to collaborate across diversity [7]. In order to cooperate to accomplish particular goals and to coordinate diverse activities, we have to seek to overcome lack of agreement, at least on some issues. As far as healthcare settings are concerned, there are at least two levels on which we cannot operate effectively and authoritatively without consensus. Agreements, laws, regulations, procedures, and consensus are necessary on the macro-economic/political level. This *macro-consensus*, however, is expressing and strengthening the fabric of societal actions, attitudes, and values, and supporting the processes of formation of *micro-consensus* at the bedside. On this second level, in daily interactions between healthcare professionals and patients, consensus is required to initiate, continue or discontinue various medical interventions.

Fundamental ethical questions concerning the formation of consensus have already been discussed in a collection of essays edited by K. Bayertz [2]. We can see how ambiguous the concept of 'consensus' really is, how its moral status is highly questionable, and how difficult it is to achieve a rationally-founded consensus. Bayertz concludes that consensus can only claim moral authority "when it is the result of a rational communicative process aimed at intersubjective understanding and a just balancing of interests" ([3], p.13). The essays in this volume focus precisely on these underlying mechanisms of consensus formation in healthcare practices.

In chapter one, Robert Veatch asks whether consensus in clinical cases is a prerequisite for good clinical practice, and whether consensus should address issues of fact or issues of ethics. There are, of course, good reasons to have expert consensus concerning some professionally legitimated standards, as in clinical pharmacology or surgery; but even those scientific agreements, Veatch argues, are not free of consensus on *normative* issues, such as mores, preferences, or ethical principles. At the bedside, the parties involved in consensus formation might be just the patient and the physician. Because the principal decisionmaker for all clinical interventions is the patient, well-informed decisionmaking, not consensus, seems to be of prime importance. To the extent that clinical decisionmaking occurs in institutional settings, other individuals and institutional policies have to be included in the process of reaching consensus. As Veatch points out, however, there is no logical need for either scientific or ethical consensus by all members of the healthcare team in a clinical case. He holds that consensus is not binding for competent patients; only in certain situations is it binding for surrogate decisionmakers. Stuart Spicker reminds us that most healthcare ethics committee discussions and resolutions represent subtle group unanimity rather than consensus. He points out that agreements reached are based on the process of successful interpersonal and social interaction, but do not represent binding moral, much less legal, authority. He sees consensus as a *process* rather than a goal, and reminds us of the roles played by 'yielders' and 'independents' during the social-psychological processes of "coherence formation."

Debates on defining the limits of solidarity and public responsibilities for healthcare are oriented and targeted towards consensus. As Henk ten Have points out, how to devise such a societal and public consensus formation process is itself an issue on which consensus has to be reached. Given the apparently simple question – How do we define *health?* – he signals three different approaches: the individual approach – relating health to autonomy and self-determination; the medical-professional approach – defining health as the absence of disease; and the community-oriented approach – defining health as the individual's ability to participate in social life. Analyzing the experiences of a special governmental Committee on Choices in Healthcare in The Netherlands, ten Have shows how a consensus emerged among the committee members, particularly in regard to the community-oriented approach; it had important consequences for the specific priorities of healthcare reform: to care, first, for those who cannot care for themselves; second, for those who can function in society since their health can be restored or maintained; and, third, for those whose lives are seriously threatened. Interestingly, the Committee did not think that it was asked to define a goal, but rather to introduce a *process* of public debate to establish the order of the priorities in a publicly-financed healthcare system.

CULTURES IN CONSENSUS FORMATION

Reformulation of the notion of 'consensus' is impossible without the recognition that different cultures and traditions are at work in forging and forming consensus. One model of consensus regards consensus as that which survives a competitive process of debate and compromise. As the outcome of fair competition, *pluralistic consensus* is an attractive model of collective agreement in a culture of interest group liberalism. A second model of consensus assumes that individuals in a given moral community have overlapping value sets. Moral dialogue is aimed at searching for those values that are common to all members of a peaceful society. However, this model of *overlapping consensus* requires that participants in deliberation de-emphasize the substantive moral views which make them different from each other. Analysis and critique of the two models of consensus, has led Jennings [10] to search for an alternative model of consensus: consensus as a dialogic and discursive activity focused on creating common grounds and developing a sense of the common good among members of a community.

Similar concerns are expressed in Part Two of this volume. Hub Zwart reviews the options available in a pluralistic society between peaceable moral deliberations and moral warfare. He presents the views of Engelhardt, Lyotard, Apel, Habermas, MacIntyre and Stout. Some of these scholars propose new rules for negotiation, contract, or discourse among moral strangers. Others, such as Spicker, suggest international and cross-cultural guidelines, such as those prepared by the Appleton group on decisions to forego medical treatment. None, however, possesses warrior-like attitudes to exclude those who use different moral languages in the process of consensus formation. Henrik Wulff discusses the relationship between cultures that employ moral, rational debate and cultures that reflect a *mores*-based ethos. These two different cultures, as well as the two different concepts of ethics, are based on 'individual rights' (represented in mainstream U.S. thinking) as well as based on 'mutual obligations' dominant in European thought, and show that ethics, including medical ethics, is actually quite culture-specific. Rational ethical discourse therefore has to take cultural and moral traditions into account, and must recognize their relative resistance to rational discourse and consensus formation.

Robert Cook-Deegan, drawing on his prior experience at the U.S. Congress' Office of Technology Assessment, reviews various models of consensus-forming bodies in U.S. public policy as well as their structures and procedures – he takes us from blue-ribbon committees to expert commissions. He called repeatedly for the establishment of a national bioethics advisory committee during the 15-year period when the U.S. remained commissionless. The 18-member U.S. National Bioethics Advisory Commission (NBAC) met formally for the first time on October 4, 1996, although President William Clinton issued the Executive Order creating NBAC on October 3, 1995. In Japan, Akio Sakai calls for the establishment of interdisciplinary ethics committees to convene at Japan's eighty medical schools, in order to engage professionals from various disciplines in the promotion of a public consensus. He points out that experts typically reflect the existing consensus and attitudes of local communities, and that there is no other way to create a new ethics than to work to achieve a broad consensus through open public discussion.

THE CLINICAL PRACTICE OF CONSENSUS FORMATION

Whatever the specific national or professional cultures at work in conflict resolution, there is a need to reflect on models of competent moral thinking, as James Drane puts it in Part Three. He reviews U.S. as well as European approaches in drafting formalized action guides by developing questionnaires for integrating patient-oriented ethics into clinical decisionmaking. The European approaches are primarily concerned with the ethical issues that underlie good clinical practice, while the U.S. position aims to protect patients' rights and autonomy; however, both arrive at the same goal – developing an applied ethics for the patient's good.

Principles of philosophy and ethics "loosen up" when applied to real-life situations, such as complex clinical decisionmaking. Bill Fulford discusses the limits of consensus formation in psychiatry; quite often we have to have clear and well-argued *dissensus* rather than consensus in treatment decisions. There are cases where scientific consensus has to be balanced by rational *dissensus* because the variability of our value judgments is integral to our individuality as human beings. Fulford calls for further philosophical research in the area of rational consensus and dissensus, for certain degrees of dissensus might not only be legitimate but a direct reflection of good clinical practice.

To explain the 'Mediterranean view' of the doctor-patient relationship, Sandro Spinsanti assumes that good medical practice and good ethical practice are related to each other like figure and ground in Gestalt theory. Good clinical practice, as he understands it, cannot be performed without recognizing the preferences of the patient and the involvement of friends and/or family, including the physician's occasional duty to protect the patient's autonomy against intrusion by the family. When consensus between patient and physicians cannot be established, medical practice becomes very complicated and problematic. Even though the Italian code of physicians' ethics, as Spinsanti shows, still offers strong arguments in favor of withholding *discouraging* information from the patient in many circumstances, consensus on treatment decisions between physician and patient is frequently not achieved. In the U.S., as Chris Hackler points out, truth-telling and consensus have become well-established practices in institutional policy and bedside decisionmaking. He discusses the struggle to reach consensus in cases of "futile" treatment, and the roles of the family and the patient in that struggle.

Specific problems in consensus formation in the context of genetic counseling are discussed by Gebhart Allert and colleagues. While non-directive counseling is morally preferred (in order to allow autonomous and decisionmaking by those counseled) directive counseling is often desired and even requested by patients and their families. Consensus between the consultant and those consulted is mandatory in regard to facts, but the authors argue that it should never be a goal in family planning decisions. These case discussions demonstrate clearly that counseling is a fine art rather than a science: How, then, can we counsel others in the postmodern world of moral strangers? – Engelhardt asks. Allert *and others* show how counselors have to explore the possibilities and impossibilities of consensus formation at various levels on a case-by-case basis.

Different family values, religious traditions, and individual principles make not only family planning, abortion, and criteria-for-death decisions difficult, but it is also difficult to shape policy for reproductive technologies, as Anne Donchin points out. In assessing the political and philosophical British debate on the moral status of modern reproductive technologies, she concludes that it would be even more difficult for European nations to develop a unified and ethically coherent policy in this or any other area of bioethical concern. Her critical reflections shed light on the ongoing debates in preparing the European Convention on Human Rights and Biomedicine. Now that the Committee of Ministers as well as the Parliamentary Assembly of the Council of Europe, after a lengthy process of negotations, have approved the text of the Convention, the member states of the Council continue to discuss the text, but always with the intention of signing it. Elaborating on existing international legislation, the Convention aims to achieve a greater unity between its members, and at safeguarding human dignity and the fundamental rights of the individual with regard to the applications of biology and medicine [6]. In 22 articles, a framework of principles and applications is set out to guide the practices of healthcare and research. Some articles are very general, others are specific and set out in rather rigid language.

Considering the difficulties and limitations of consensus formation in the daily practice of healthcare ethics, the Convention can be viewed as a major international effort to obtain consensus between European member states, and can equally be regarded as an epoch-making experiment. Although the text is still under consideration by national parliaments, and the final outcome of the political decisionmaking not clear, fundamental questions can be raised concerning the attempt to draft a document that reflects shared European premises for healthcare ethics. But, one can ask, do we really need a uniform European convention framed around bioethical principles? Does a common European value-framework support such a convention? We should also note here the *Janus-face* of consensus formation: the convention, drawing on a nucleus of consensus, will at the same time create other nuclei of consensus. Nonetheless, this process of consensus formation will be interesting to follow. Will the European culture of rational moral discourse better support a convention that would stress the rights, responsibilities, and autonomy of individual and familial judgment over societal and political regulations and rules? Should a European convention be the product of an open and free consensus formation processes among European citizens rather than the result of intergovernmental negotiations in ethics diplomacy? There seems to be no consensus on how to answer these questions, either among the Europeans or their elected politicians and bureaucrats.

WHERE DO WE GO FROM HERE?

Finally, the question can be asked again: Is there a need for consensus in bioethics? The simple and noncontroversial answer given by all the contributions to this volume is "yes" – we need consensus; moreover, we have already achieved consensus at various levels. The authors also agree that the "depth" of consensus depends on the situation and on the level of decisionmaking. Healthcare professionals have special ethical duties toward each patient irrespective and quite independent of the patient's legal rights. Indeed, the very practice of medicine is presently undergoing a metamorphosis. Essential parts of this metamorphosis are the means and goals we derived from the formation of consensus on various levels in healthcare and clinical practice.

Given the range of human need in general, and of the needs of the sick, the poor, and those who are suffering (together with limited financial and human resources), we need to achieve societal consensus on priorities, especially in allocating resources for healthcare. There are different cultures in which a macro-allocational consensus has been established, either by elitist, expert agreements, political processes, market or "managed" market forces, national and local committees, public discourse, or a combination of these. All cultures seem to pay homage to the principle of solidarity at least when determining the "basic healthcare package" that guarantees access to healthcare at affordable cost and reasonable quality.

In pluralistic societies, the problems of how much societal and cultural consensus is needed to deal responsibly with the "edges" of individual human life, and how far these issues should be left to individual conscience may never be finally resolved. In more technical terms, we shall continue to face the problem of whether the principle of *subsidiarity* can be transported from social ethics to bioethics in order to reduce societal and religious conflicts, where individual value systems do not concur. Where we cannot reach consensus on issues of *content*, we shall have to try even harder to pursue peaceable methods of conflict resolution and to recognize that we have the right to disagree on some values as we agree on others. Whenever we do not agree on the moral assessment of facts or actions, the formation of consensus in dealing with dissent will not be easy. But rational reflections and agreements on process promise greater success and less conflict in the pragmatic quest for peaceable solutions than do their absence, even though the solutions thereby reached might not convince everyone. A Japanese proverb might offer some hope concerning the open-ended processes of consensus formation, not only in Europe and not only in bioethics: "On the road you need a companion. In life you need kindness."

REFERENCES

1. Bauman, Z.: 1995, *Life in Fragments: Essays in Postmodern Morality*, Blackwell Publishers, Oxford, U.K.
2. Bayertz, K. (ed.): 1994, *The Concept of Moral Consensus. The case of technological interventions in human reproduction*, Kluwer Academic Publishers, Dordrecht, The Netherlands.
3. Bayertz, K.: 1994, 'Introduction. Moral consensus as a social and philosophical problem', see ([2], pp. 1-15).
4. Byk, C.: 1997, 'A proposed draft protocol for the European Convention on Biomedicine relating to research on the human embryo and fetus', *Journal of Medical Ethics* 23, 32-37.
5. Caws, P.: 1991, 'Committees and consensus: How many heads are better than one?', *The Journal of Medicine and Philosophy* 16, 375-391.
6. Directorate of Legal Affairs: 1996, *Convention on human rights and biomedicine*, Council of Europe, Strasbourg, France.
7. Engelhardt, H.T., Jr.: 1994, 'A Skeptical Postscript: Some concluding reflections on consensus', see ([2], pp. 235-240).
8. European Bioethics Convention: Draft convention for the protection of human rights and dignity of the human being with regard to the application of biology and medicine: 1994, *International Journal of Bioethics* 5(2), 131-138.
9. Government Committee on Choices in Health Care: 1992, *Choices in Health Care*, Ministry of Welfare, Health and Cultural Affairs, Rijswijk, The Netherlands.
10. Jennings, B.: 1991, 'Possibilities of consensus: Toward democratic moral discourse', *The Journal of Medicine and Philosophy* 16, 447-463.
11. Rogers, A. and Durand de Bousingen, D.: 1995, *Bioethics in Europe*, Council of Europe Press, Strasbourg, France.
12. Spinoza, B.: 1670, *Tractatus Theologico-Politicus*, Henricus Künrath, Hamburg, Germany.
13. de Wachter, M.A.M.: 1997, 'The European Convention on Bioethics', *Hastings Center Report* 27 (1), 13-23.

PART I

CONSENSUS FORMATION AND HEALTHCARE

ROBERT M. VEATCH

ETHICAL CONSENSUS FORMATION IN CLINICAL CASES

It is widely believed that ethical consensus in controversial issues is an ideal worthy of pursuit. This is no less true in healthcare than in other spheres. This quest for consensus has included a desire for agreement on both the beliefs about the technical facts and the normative assessment of what ought to be done. At the public policy level health planners hope to achieve consensus around healthcare plans, standards of practice, and collective judgments of safety and efficacy. This has given rise to devices for the generation of consensus. In the United States these have included federal panels such as the National Commission for the Protection of Human Subjects [14], the President's Commission for the Study of Ethical Problems in Medicine and Biomedical and Behavioral Research [17], the federal executive branch Ethics Advisory Board [6], and the Congressional Biomedical Ethics Board [10]. At the National Institutes of Health the goal of consensus has produced an ongoing public policy mechanism involving panels of experts meeting as Consensus Conferences [15].

Consensus is often believed to be even more important in the clinic. Clinicians often work under the assumption that clinical cases, no matter how complex, generally ought to generate consensus–at least as an ideal. Many of the concepts of the clinic imply that such consensus is normative. Both in professional practice and in law, a standard of practice is believed to be significant for defining not only what is done, but what ought to be done. In many legal systems that standard of practice can serve as the basis for legal action involving malpractice, negligence, and other matters.

Many of the conceptual tools of clinical medicine involve ideas implying consensus. For example, clinical pharmacology makes frequent use of the notion of a "treatment of choice" or a "medically indicated treatment." Both of these imply that it is believed to be possible to identify, presumably on the basis of some professionally legitimated standard, diagnoses and prognoses under which certain treatments are "medically correct." Not only are they believed to be medically correct; they can actually be identified as the best treatments, that is, treatments of choice.

H.A.M.J. ten Have and H.-M. Sass (eds.), Consensus Formation in Healthcare Ethics, 17–34.

These clearly involve normative, as well as technical, agreement. They are based on evaluations of how alternative treatment candidates stack up according to some agreed upon system of evaluation as well as technical agreement about exactly what the patient's condition is and what the effects of alternative interventions are likely to be.

Recently, in order to further the process of consensus formation in clinical cases, mechanisms have emerged that will help achieve the goal. Peer review groups help reach agreement on more technical matters. Often, however, these peer review groups move considerably beyond the merely technical. They render collective wisdom on whether a treatment offers expected benefits exceeding the expected risks. They are particularly active in assessing whether interventions produce enough expected net benefit to be worth providing. Peer review has been active in criticizing tonsillectomy, hysterectomy, and other problematic surgical procedures. They have been used to establish formularies of pharmaceuticals considered potentially useful.

These peer review judgments clearly extend well beyond any consensus that could be based solely on scientific facts alone. They are not limited to assessments of what the effects of various surgical and pharmaceutical options are. They include explicit judgments about whether the expected outcomes are worth pursuing. Still, many practitioners probably are unaware that they are pursuing normative as well as scientific consensus.

In some cases, however, the ethical dimension of the consensus is particularly vivid. The cases involve blatant ethical judgments that somehow are perceived as more conspicuously normative. Moreover, the normative assessment is understood to be based on ethical rather than other kinds of normative judgments, say those involving aesthetics, mores, or preferences. When ethical matters appear to be at stake, recent clinical practice has supported the use of ethics committees as a way of striving for moral consensus [12,2,3,25]. It may be, however, that we have not asked with adequate seriousness what the significance is of such consensus. Our belief in the importance of consensus in clinical cases may be without foundation.

I. THE IMPORTANCE OF CONSENSUS

Some of the advocates of consensus are willing to acknowledge that in matters of public policy, especially in a complex, pluralist society, consensus may not always be necessary and may, in fact, be impossible to achieve. They may still hold on to the ideal of consensus regarding the clinical case. They may still believe that the members of the healthcare team ought to come to some working agreement on the correct course for the patient. They may, at least implicitly, hold to such concepts as the treatment of choice, the medically indicated treatment, and the standard practice, implying that consensus is both possible and normatively important.

In assigning this relative value to clinical consensus as compared with public policy consensus they may have things just backwards. For some matters of public policy, there simply has to be agreement. Heroin either is or is not such a dangerous drug that it should be outlawed. On the question of making certain medical practices illegal, there can be no pluralism. Participation of the clinician in an active killing of a suffering patient may be legalized or made illegal. It must either be tolerated or proscribed. The law, however, cannot avoid some coming to terms. There may be many practices in medicine, which, as a matter of public policy, may lend themselves to more individualized discretion. Patients and physicians may choose to participate in aggressive cancer chemotherapy or palliative hospice care. They cannot, however, avoid a decision on whether "supportive care only" is within the range of the acceptable. In both public and private health insurance there must be agreement on whether a healthcare service is covered. Even if patients are free to refuse covered services, there must be a consensus on what services are to be made available under insurance programs. Public policy requires some level of agreement. This is true both on matters of scientific fact and on matters of normative judgment. It is true on matters ethical (abortion, mercy killing, and the definition of death) as well as matters technical.

At the clinical level, however, there may be less real need for consensus. Individual patients and individual clinicians can function quite adequately whether or not other onlookers concur in the judgments of the individuals involved. To the extent that clinical care is rendered within some institutional setting such as a hospital or nursing home, then the individuals making clinical judgments may be obliged to conform to the range of acceptable practices and policies of the institution. That, however, is to retreat to the need for some range of agreement at the policy level, albeit policy at the level of the institution. By contrast, there is no logical need for either scientific or ethical consensus by all members of the healthcare team in a clinical case. In fact, no one need concur in the judgment except those directly acting in the situation. In the simplest case only the patient need endorse certain decisions the patient makes (such as the decision to leave a hospital or forgo medical treatment). Even if there are actions requiring cooperation of a health professional, only the individual care giver need concur in the decision.

Consensus as an Epistemological Device

Underlying this concern over consensus is a question of what function consensus ought to serve in clinical cases. At the sociological level consensus may serve to provide an atmosphere of cooperation. Physicians working with the consensus of the members of the healthcare team, the patient, and other members of the lay team, can work with greater confidence of cooperation and less fear of a law suit. However, consensus serves an important moral function as well. Consensus is, in some sense, an epistemological device. Surely, one of the reasons why there is such interest in consensus is the belief that if consensus exists there are better grounds for believing in the correctness of the judgment. This is true both regarding matters of scientific fact and normative judgment. Thus both elements of a clinical decision–the factual and the evaluative–are thought to be served by consensus.

At the same time it is one of the fundamental insights of Western epistemology that consensus does not guarantee correctness. Since the corporate condemnation of Socrates in the Greek tradition and Jesus in the Christian, we have been impressed by the possibility of error by the majority and the tyranny that can result if actions are dictated by it [13]. It might be argued that these are not examples of full consensus because the eventual heros of both events explicitly were not part of the consensus. Even so, it seems clear that if a full consensus did exist–a consensus involving literally every member of the community–it still would not guarantee correctness.

This epistemological conclusion holds for both science and normative evaluation. The consensus of peers in a scientific assessment is of great interest, but it would be a mistake to assign definitive correctness to a peer review. Likewise, normative assessments by consensus need not be definitive. Some nonmoral evaluations, such as personal preferences, may be thought to be immune to consensus because they have no objective truth value. Others, such as moral evaluations, are normally thought to have an objective referent, but their correctness is still not known definitively by consensus.

Contract Theory and Consensus

In spite of the disjunction between consensus and objective correctness, consensus seems important. It is even important as an epistemological device, at least in some epistemological theories. Contract theory is an example in contemporary ethics of a system in which consensus plays an important role. Rawlsian contract theory identifies the fundamental moral principles of a society by asking what people under certain carefully specified conditions would agree to as the principles for structuring social practices [18]. For Rawls "pure procedural justice" derives from agreements contractors would reach under the veil of ignorance, that is by parties who were masked as to their specific positions and interests in the society (see [18], p. 85). At least under the conditions of the hypothetical contract, consensus can produce the ethically correct principles, by definition.

This analytical linking of consensus and correctness may be unacceptable to others such as those standing in the major Western monotheistic religious traditions. They, presumably, relate correctness in both science and morality to an objective standard external to the agreement of humans, even hypothetical humans behind the veil of ignorance. However, some religious ethicists place great emphasis on an ethical method that is quite similar to contract theory. In covenant theory, ethics is a matter of promising, of covenanting. Often the covenants are between a human community and a deity, but sometimes they are among humans. What is important for our purposes is that covenant theory can also use something akin to consensus as an epistemological device.

How do finite, fallible humans know the divine will in religious ethics? In some such systems, the best that can be done is to gather together the members of the community and have them attempt to discern the divine order. In some of these traditions, this means discerning the natural law. Agreement by the community of observers is the best we have to go on. In the Old Testament book of Nehemiah, after the Israelites returned from the exile, they needed to re-establish the covenant (Neh. 10:28-31). In order to establish the content of the covenant, there was an assembly of all the people–the ordinary laborers as well as the priests and Levites. Through the collective wisdom of the group, the content of the covenant was discovered. It may not be perfect knowledge, but it is the best that finite humans can do. The contract method in religious ethics is a method for discovering the truth rather than inventing it. Some secular philosophical methods set out on the same task of discovery. They rely on reason, empirical observation, or an "ideal observer" [7] to discern a preexisting moral truth. To the extent that they rely on a convergence of many persons reasoning or observing together, they also use consensus as an epistemological device. Here consensus is linked to the truth synthetically rather than analytically, but it is linked nonetheless.

Thus the consensus of the community emerges as an epistemological device–a way of knowing–both in secular philosophical method and in religious ethics. For those who consider truth to be objective and discoverable as well as for some who consider it invented, consensus is a critical epistemological device, not infallible, but about the best humans can do.

Shortcomings in Clinical Consensus Mechanisms

In order for consensus to work as an epistemological device certain criteria have to be met. For Rawls the criteria are summarized as the veil of ignorance [18]. The correct moral principles are those that hypothetical contractors behind such a veil would agree to. For real people, the task is to strive to come as close as possible to the conditions behind the veil: to try to avoid being influenced by idiosyncratic personal preferences and knowledge of one's position in the social system. Likewise, for other secular systems the convergence of a consensus is a useful way to approximate pre-existing moral truth just as it is in religious ethics. What does this mean for the pursuit of consensus in clinical ethical cases? Peer review and other collective judgments seem to be built on the device of group consensus as a way of improving the probability that a correct answer is obtained. Insofar as what is at stake is an ethical judgment about how a patient or healthcare professional ought to behave or be treated, it is fashionable to seek an ethical consensus either among the members of the healthcare team, among professional colleagues, or among the members of a hospital ethics committee. At a larger level similar consensus is sought among the members of national and international professional organizations. Limiting our attention to specific clinical cases, we will primarily be interested in the pursuit of consensus at the local level among clinical colleagues, members of the healthcare team, or a hospital ethics committee. Can consensus serve as an epistemological device for improving the quality of clinical ethical judgments?

It would be strange if sometimes it did not. Surely, many heads may be better than one. Individual practitioners may make idiosyncratic judgments. They may carelessly omit from consideration critical facts or moral considerations. I think we need to be equally impressed, however, with the shortcomings in the clinical consensus mechanisms. Insofar as the argument for consensus as an epistemological device works for establishing an increased probability of a correct judgment, most of the theories we have examined rely on mechanisms for neutralizing special points of view. They make members of the group blind to their specific interests and position, they make those contributing to the consensus an all-inclusive group (as in Nehemiah), or they make the observers "ideal," that is all-knowing and "omnipercipient," capable of empathizing with all parties affected by a decision. The consensus is useful only to the extent that it is not biased in some way.

The Atypical Views of Clinicians

The first problem we encounter if we apply the consensus mechanism to clinical cases is that the group through which a consensus is likely to emerge is dramatically different from the ideal of a group behind a veil of ignorance. If the consensus we have in mind is the consensus of physician-clinicians at a particular health facility, we should immediately perceive the shortcomings. They are all physicians; they are all trained to deal with a specific kind of expertise. They are socialized into a set of values and a historical ethic. Some clinicians go so far as to claim that there is an ethic "internal to the practice of medicine" ([11,16,1,5]). They believe that by analysis of the very concept of medicine, values and goals of medicine can be discerned. Historically some clinicians have held that only clinicians are capable of this knowledge. They, thereby, derive such goods as the preservation of life, the curing of disease, the relief of suffering, and the preservation of health. They also believe they can know what the proper priorities are when these so-called medical goods come into conflict. It seems clear that other persons outside of the clinical consensus would identify different goods that could be achieved with the medical professional's talent and that they would arrange those already identified with radically different priorities. There seems to be nothing in one's training as a physician that would give one expertise in identifying these norms and ranking their priority.

The problem is even more severe. Medical goods are often in tension with nonmedical goods. Preserving a patient's life or relieving suffering may conflict with using one's time, energy, and other resources for other goods: pursuing goods of the family, the religious life, one's occupation, or personal pleasure. Insofar as clinical professionals have chosen medicine as their career, it seems clear that they have come to place abnormally high value on medicine. Not only did they value medicine's goods abnormally when they entered the profession; they have been socialized during their years in the profession to develop further their unique perspective. It seems clear that clinicians ought to be expected to converge around the unique beliefs and values they hold. Their collective *Weltanschauung* ought to show up in their consensus.

This might lead us to seek a broader consensus. Especially on moral matters this is often done. This is the moral reason consensus might be sought among all the members of the healthcare team. If nurses, social workers, and other care providers are brought into the consensus, surely the quality of that consensus improves as more perspectives are added. The result is not only greater cooperation among the team, but a more solid basis for believing that the correct decision is reached.

Even then, however, one must be impressed by the great distance of such a group from the ideal of hypothetical contractors. All health professionals share a common Hippocratic heritage, including all its evils. They share the commitment to the clinical perspective, to caring for the patient, and to medicine. As such they represent as very narrow range of observers. Their consensus is surely far from that that would emerge if all economic, social, racial, religious, cultural, gender, and age perspectives contributed.

Moving to a hospital ethics committee broadens the perspective still further. The typical hospital ethics committee includes clinicians who have devoted unusual amounts of time to studying the ethics of cases. It normally includes nonclinical perspectives as well. It may include a philosopher, priest, lawyer, business person, or homemaker. It may include former patients or relatives. To the extent is does, it produces a consensus that is much better than any that could derive from mere clinicians.

Still, the gap between the typical ethics committee and the ideal Rawlsian group of contractors is gigantic. Virtually every hospital ethics committee is dominated by physicians. The chairman is typically a physician. A physician and nurse co-chair in some of the more enlightened committees in the United States. Still, they exist under the auspices of the hospital. At most nonclinicians are a minority. The consensus that emerges ought to be distorted by the clinical perspective.

Other distortions may creep in as well. The literature of social psychology documents a phenomenon of group decisionmaking known as "risk-shift" [8]. When groups make collective decisions, the result is not necessarily the "average" pre-existing view of the members of the group. Each person's judgment may shift away from his or her previously held personal view. In collective decisions pertaining to risk-taking, social psychologists have documented that the resulting collective decision is predictably more in favor of taking risks than is the average of the pre-existing individual positions of the group members.

The reasons for this are complex. Perhaps each person feels less responsibility for the resulting judgment than he or she would making it individually. They are each willing to take more of a chance when the decision is attributed to the whole group. Moreover, in some groups the variation in views is not normally distributed. In many groups studied, the pro-risk-taking view dominated. If each speaker states his or her view, more pro-risk-taking statements are heard by each member of the group, possibly leading to a shift toward the predominating statements. Although the studies did not include groups dominated by anti-risk takers, presumably if there were such a group, the shift could be in the opposite direction. What is important, for our purposes, is that the result of the group decisionmaking can not be presumed to be the average of the individual judgments and the shift takes place in the direction of the dominant members.

This is important for hospital ethics committees because the committees are normally dominated by the clinical perspective of one or more physicians even if a wide range of other perspectives is represented. The consensus that emerges might be given undue weight as representing a broad consensus, when, in fact, it still represents the traditional narrow consensus of the physician-clinicians.

Variations from One Clinical Consensus to Another

The value of a clinical consensus is even more questionable when one realizes that there is tremendous variation from one clinical consensus to another over apparently similar cases. We know that two groups of health professionals in the same or a different hospital are likely to reach different consensuses on the same problem. Different ethics committees will evaluate the same case differently. The empirical data are not yet in to provide objective support for this conclusion, but anecdotal evidence makes this clear. The documentation is available in the United States for a rather similar kind of committee–the institutional review board for approving research on human subjects. When different committees are given protocols, the results they get are dramatically different ([9,4]). There are good theoretical reasons why hospital ethics committee consensuses should vary even more than institutional review boards. IRBs in the United States must conform to a standard set of federal regulations [19]. Their members must conform to certain federal standards regarding composition. Individual members of the IRBs must be approved. Decisions that are out of line can be reviewed by a federal government agency, which has strong sanctions it can apply including suspending federal funding to the institution that does not conform.

By contrast hospital ethics committees (and other groups generating clinical consensus such as peer review bodies and healthcare teams) are strictly local entities. With few exceptions [20] local boards do not need to conform to any standardized rules. They have no public mandate. They perform strictly local functions and are appointed by local authorities.

All healthcare facilities must operate based on some implicit or explicit ideology. It may be the ideology of a religious sponsor, a private, voluntary association, or a commercial commitment. The sponsors can be expected to have differences in their ideologies. Some will be dramatic and visible such as a hospice or an aggressive cancer research orientation; an orthodox Jewish commitment or a feminist perspective. Others will be less conspicuous, but no less important in shaping the functioning of the facility. The administrators of the facility have a moral (and perhaps a legal) duty to operate their institution based on the ideology of the sponsors. It would be irresponsible of them to fail to appoint their staff committees, peer review groups, and ethics committees without some eye toward their institution's basic philosophy.

If this is true, then the ethics committee or other clinical consensus formation mechanisms of the local institution ought to be structured so as to reflect that ideological commitment. Of course, a sophisticated administrator will not appoint all members of the consensus-forming group to reflect precisely the ideology of the institution. For the institution's own good and to generate greater insight some members of the group will deviate from the institution's ideology. A mix of religious, secular, political, and ethical perspectives will appear, but the dominant and median view ought to be approximately that of the institution's sponsor. To do otherwise would be strange. A Catholic hospital ought to have a substantial Catholic presence; a Marxist hospital, a Marxist gestalt, and so forth.

If this is true then not only will one healthcare team or consensus forming body at a local institution vary randomly from another at that institution. Various consensus-forming bodies reflect systematic biases generated by their common commitment to the unique perspective of the healthcare professions, but still also predictably vary as a function of the ideology of the sponsoring institution. In the end the consensus generated ought to reflect in part the system of belief and value of the healthcare professions as a whole, in part the special idiosyncrasies of the individual practitioners, and in part the special idiosyncrasies of the sponsoring institution in which the healthcare is delivered. The result should have little in common with what would emerge as a consensus of hypothetical contractors behind a veil of ignorance or of a highly diverse body functioning according to democratic theory. The consensus generated in a clinical setting ought not to be worth very much.

II. IS CONSENSUS A LEGITIMATE GOAL IN CLINICAL CASES?

All of this implies that the consensus of a more ideal group would be significant in the first place. Yet, as I have suggested earlier, in the clinical setting (as opposed to the policy setting), a consensus may not be urgent or important. Even if we could obtain what I will call a "valid" consensus, that is one that approximates what would emerge from rational contractors behind a veil of ignorance, it is not clear how meaningful it would be.

It will depend on the status of the patient. I will, from this point on, assume that the patient, if competent, or the surrogate for the patient, if the patient is incompetent, is to be considered the legitimate primary decisionmaker in all clinical cases [24]. By primary decisionmaker, I mean the patient or patient surrogate should be given the major responsibility for the primary decisions. Of course, other actors will have secondary decisions to make that are important. If the patient chooses a controversial course of action, the physician will have the moral responsibility of deciding whether to remain part of the lay-professional relation. He may choose to withdraw from the covenant with the patient. The physician will also have to decide what information is transmitted to the patient from the infinite array of items that could be. Other members of the healthcare team also have important decisions to make. The nurse will have to decide which nursing options to present to the patient and decide whether to remain part of the team if the patient chooses a controversial course. The primary decision, however, must rest with the patient or patient surrogate.

The competent patient or patient surrogate would, of course, be interested in a correct, objective assessment of both the scientific facts and ethical evaluations, to the extent that these can be obtained. Only the most stubborn lay person would turn his or her back on such wisdom. Even if it is available, however, it need not be definitive, at least in any ethical theory that includes a principle of autonomy.

The Competent Patient

In the case of the competent patient, it is possible that judgment would not be swayed even by hypothetically definitively correct moral advice from the consensus group. He or she may choose to act unethically and, according to the principle of autonomy should, in some cases, be permitted to do so (at least if other ethical principles are not violated such as the principle of justice or harm to others).

In the real world in which the consensus from the clinically dominated group is far from definitive, the competent patient may also choose to deviate from that consensus for what to the patient are good moral reasons. The patient may choose to get moral knowledge from some other source. This would particularly be true in the case of a patient from a cultural group different from that of the group generating the consensus. Jewish patients ought not be swayed by the consensus of the ethics committee of a Catholic or a secular hospital.

If the purpose of the consensus-forming group is to advise the primary decisionmaker (the patient) about the ethical options, at the very least the primary decisionmaker ought to be present for the deliberations of the group. To meet without the permission of the patient and without an invitation to the patient to be present is unethical and foolish. It may also be illegal.

It is foolish because the easiest way to help the decisionmaker understand the range of moral options is to let the decisionmaker observe and actively participate in the deliberation of the group. It is unethical to the extent that it implies a less than active involvement in the decision by the patient. It is illegal (at least if the patient has not given permission for the group to meet) in that it involves transferring personal, private, privileged information about the patient without the patient knowing and approving. Confidential information belongs to the patient and cannot be given to another without the patient's permission.

Critics might argue that patient information is transferred routinely within a hospital to other professionals or to hospital committees. This is different, however. Confidential information is transferred legitimately either when required by law (such as to certain peer review committees) or when one has the consent, presumed or explicit, of the patient. Casual consultations with colleagues normally can take place with the presumed consent of the patient. He could not plausibly object.

Transfer of information to an ethics committee or other body designed to generate moral consensus, however, is another matter. It is not authorized by law and one cannot presume the patient will consent. He should not, if he has other sources of moral counsel and believes he does not share the value perspective dominating the committee. The committee could cause him trouble. It could take steps (legal and clinical) to block the patient's plan of action. In some cases the rational and prudent patient would refuse to give permission.

Even if he or she does give permission and participate in the deliberations of the consensus-forming body, there may be good reasons why the patient would not accept the consensus so formed. The patient may evaluate the situation differently. She may use a different ethical theory, quantify the benefits and harms of options differently, or even evaluate certain beliefs about the facts differently. To the extent the patient is to be given free reign as an autonomous decisionmaker, the consensus of real clinical groups need not be definitive. The prudent patient will probably want to participate in the deliberations of the consensus-forming group in most cases. He should be present and should be dominant in the deliberations. He should play the role of "captain of the team." A prudent patient will often want to hear even from groups with which he knows he disagrees. It is often far more valuable to hear the arguments for positions one does not already hold. The consensus of such a group, however, has no weight against the reflective judgment of the competent patient.

The Incompetent Patient

For the surrogate for an incompetent patient, the matter is a bit more complex. If the surrogate is providing a substituted judgment based on the patient's wishes expressed while competent, the same conclusion applies, provided the surrogate reasonably reflects the patients' views. But can we give the surrogate for the incompetent who has never expressed relevant wishes while competent such free reign?

Not quite. The bonded surrogate for such a patient–family member or surrogate designated by a durable power of attorney–should try to do what is best for the patient. He or she would reasonably want to listen especially to groups who can provide alternative positions to his or her own. The surrogate's judgment of what is best, however, may not conform to the consensus. I have suggested reasons why in some cases it ought not to conform to the consensus. In those cases within certain limits the surrogate's position should take precedence over the clinical consensus. The limits are worth a word. I have elsewhere developed the concept of a "limit of reasonableness" on bonded surrogate decisions ([22,23]). Any decision by a guardian that is reasonable should be accepted by clinicians. There are some judgments of what is in the interests of the patient that are simply beyond reason. Some mechanism should be available to overturn such decisions.

III. TWO SPECIAL CASES

If generally groups that form clinical consensus should not not be considered authoritative there may be two cases deserving special consideration. First, a small group of patients are particularly vulnerable. Those who are incompetent and have no bonded surrogate must have some person or group to function as decisionmaker. Could we rely on the clinically generated consensus process to make clinical decisions for such patients?

Second, I have just suggested that in some extreme cases the bonded surrogate of the incompetent patient may go too far beyond reason to let his or her decision stand unchallenged. Could we rely on the consensus of a clinical group to determine when the bonded surrogate has gone too far as well as function as the primary decisionmaker for the incompetent without a bonded surrogate?

Even in these cases there are real problems in relying on the clinical consensus to be definitive. I have given arguments why the clinical consensus ought to be unstable, show random variation, and manifest systematic biases. It seems particularly unfair to subject these vulnerable groups of incompetent patients to such unstable clinical consensus. If I am correct about the deviation of clinical consensus from the hypothetical contractor and about the variation from one group consensus to another, then a private group of citizens, even if it happens to be clinicians, should not have the authority to override a surrogate any more than it can override a competent patient. Some publicly legitimated due process will be required such as court review [21]. At most a clinical consensus that the surrogate is possibly beyond reason should be used to trigger a more public, legitimated review. Even that seems problematic if we accept the claim that patient permission is needed to initiate a consensus formation process. To turn on the patient and initiate public review proceedings would seem like a double-cross of the patient who was authorizing committee assistance. Such practices would tend to discourage patients and surrogates from authorizing committee proceedings in the first place.

If a clinical group is to function as the de facto surrogate for a guardianless incompetent patient or is to have authority for overturning the decision of a bonded surrogate who goes beyond reason, some major changes would be in order. First, these private groups would have to be carefully defined and given a public mandate for this new role. At present no private clinical individual or group has any authority to take on surrogacy for incompetents or overrule existing surrogates. A law would have to be passed establishing this role.

Second, it is hard to imagine that such clinical groups could operate outside the public nexus. Formal rules of proceeding would have to be adopted such as exist for IRBs. Due process rules would have to be established. The composition of the group generating the consensus would have to be defined. Individual group members would probably have to receive public approval as they do for IRBs. It is hard to see why society would want this responsibility to rest with a clinical group rather than one that more closely replicated the hypothetical contractors. It seems hard to believe that such groups should be based in a particular healthcare facility such as ethics committees now are. Such groups would have to resemble more closely public agencies (such as child welfare agencies), which have the legal authority to take custody and make decisions in the best interest of incompetents when their familial surrogates exceed the limits of reason. Short of this even if one rejects the primacy of the principle of autonomy, it is irrational to rely on clinical consensus.

IV. CONCLUSION

In short, clinical consensus formation is philosophically problematic. Whether the consensus is formed for issues of fact or issues of ethics, whether the consensus is reached among clinicians, healthcare teams, or special ethics committees, there are good reasons to expect in theory that the consensus reached will not have great epistemological value at getting to any underlying truth that may exist. While clinical consensus would be interesting to prudent decisionmakers, it normally should not be seen as definitive morally, legally, or scientifically.

Even if we develop a consensus formation process that approximates the standards of the hypothetical contractors, it is still unclear what the significance is of a clinical consensus. For substantially autonomous competent patients, they would surely want to be informed of such a consensus, but need not feel bound by it. To the extent that the consensus falls short of the hypothetical contractor ideal, the patient surely should not be bound by it. Likewise, for incompetent patients, the clinical consensus is informative, but not definitive. Only if there were major changes in current procedures including legal action legitimating the consensus process and changes to broaden the consensus to a public process would such a consensus be definitive even in the problematic cases of the guardianless incompetent and the bonded surrogate who exceeds the limits of reason. Clinical consensus formation in ethics as well as in science is a dangerous business and should be used only with the greatest of caution.

Kennedy Institute of Ethics
Georgetown University
Washington D.C., USA

BIBLIOGRAPHY

1. Cassell, E.J.: 1985, *The Healer's Art*, MIT Press, Cambridge, Massachusetts.

2. Cohen, C.B.: 1982, 'Interdisciplinary Consultation on the Care of the Critically Ill and Dying: The Role of One Hospital Ethics Committee', *Critical Care Medicine* (10 November), 776-784.
3. Cranford, R.E. and Doudera A.E.: 1984, 'The Emergence of Institutional Ethics Committees', in R.E. Cranford and A.E. Doudera (eds.), *Institutional Ethics Committees and Health Care Decision Making*, American Society of Law & Medicine, Ann Arbor, Michigan, pp. 5-21.
4. DuVal, B.S., Jr: 1979, 'The Human Subjects Protection Committee: An Experiment in Decentralized Federal Regulation', *American Bar Foundation Research Journal*, 571-688.
5. Dyer, A.R.: 1985, 'Virtue and Medicine: A Physician's Analysis', in E.E. Shelp (ed.), *Virtue and Medicine: Exploration in the Character of Medicine*, D. Reidel Publishing Co., Dordrecht, Holland, pp. 223-35.
6. Ethics Advisory Board: 1979, *Report and Conclusions: HEW Support of Research Involving Human In Vitro Fertilization and Embryo Transfer*, U.S. Government Printing Office, Washington D.C.
7. Firth, R.: 1952, 'Ethical Absolutism and the Ideal Observer Theory', *Philosophy and Phenomonological Research* **12**, 317-45.
8. Fraser, C. *et al.*: 1971, 'Risky Shifts, Cautious Shifts and Group Polarization', *European Journal of Social Psychology* **1**, 7-30.
9. Gray, B.H.: 1975, *Human Subjects in Medical Experimentation*, Wiley-Interscience, New York.
10. Health Research Extension Act of 1985: 1985, P.L. 99-158, Statutes at Large, 99, 820-86 (Title IV, Sections 1-12).
11. Kass, L.R.: 1985, *Toward A More Natural Science*, The Free Press, New York.
12. Massachusetts General Hospital Clinical Care Committee: 1976, 'Optimum Care for Hopelessly Ill Patients', *The New England Journal of Medicine* **295** (August 12), 362-364.
13. Moreno, J.D.: 1991, 'Consensus, Contracts, and Committees', *The Journal of Medicine and Philosophy* **16**, 393-408.
14. National Commission for the Protection of Human Subjects of Biomedical and Behavioral Research: 1978, *The Belmont Report: Ethical Principles and Guidelines for the Protection of Human Subjects of Research*, U.S. Government Printing Office, Washington, D.C.
15. National Institutes of Health: 1981, *Consensus Development Conference Summaries*, volume 3, U.S. Government Pnnting Office, Washington D.C.
16. Pellegrino, E.D. and Thomasma, D.C.: 1988, *For the Patient's Good: The Restoration of Beneficence in Health Care*, Oxford University Press, New York.
17. President's Commission for the Study of Ethical Problems in Medicine and Biomedical and Behavioral Research: 1983, *Summing Up: Final Report on Studies of the Ethical and Legal Problems in Medicine and Biomedical and Behavioral Research*, U.S. Government Printing Office, Washington D.C.
18. Rawls, J.: 1971, *A Theory of Justice*, Harvard University Press, Cambridge, Massachusetts, pp. 136-42.
19. U.S. Department of Health and Human Services: 1981, 'Final Regulations Amending Basic HHS Policy for the Protection of Human Research Subjects: Final Rule: 45 CFR 46', *Federal Register: Rules and Regulations* **46**, No. 16, January 26, pp. 8366-8392.
20. U.S. Department of Health and Human Services: 1985, 'Infant Care Review Committees-Model Guidelines,' *Federal Register: Notices* **50**, No. 72, April 15, pp. 14893-14901.
21. Veatch, R.M.: 1983, 'Ethics Committees: Are They Legitimate?' *Ethics Committee Newsletter* **1** (November), 1.

22. Veatch, R.M.: 1984, 'Limits of Guardian Treatment Refusal: A Reasonableness Standard', *American Journal of Law and Medicine* **9** (No. 4, Winter), pp. 427-468.
23. Veatch, R.M.: 1989, *Death, Dying, and the Biological Revolution*, Revised Edition, Yale University Press, New Haven, Connecticut.
24. Veatch, R.M.: 1989, 'Clinical Ethics, Applied Ethics, and Theory' in B. Hoffmaster, B. Freedman, and G. Fraser (eds.), *Clinical Ethics: Theory and Practice*, Humana Press, Clifton, N.J., pp. 7-25.
25. Weinstein, B.D. (ed.): 1986, *Ethics in the Hospital Setting: Proceedings of the West Virginia Conference on Hospital Ethics Committees*, The West Virginia University Press, Morgantown, WV.

STUART F. SPICKER

THE PROCESS OF COHERENCE FORMATION IN HEALTHCARE ETHICS COMMITTEES: THE CONSENSUS PROCESS, SOCIAL AUTHORITY AND ETHICAL JUDGMENTS

I. THE QUEST FOR UNANIMITY AND MUTUAL AGREEMENT

In May 1988, thirty-three delegates from ten countries convened in Appleton, Wisconsin in order to establish international guidelines for treatment abatement decisions and procedures for hospitalized patients. These guidelines, directed toward the formulation of healthcare policies were eventually published under the title "The Appleton Consensus." During the convention the delegates even discussed the process of searching for consensus ([15], p. 129), but in a note the author of the report remarks that "Despite the wide variety of medical cultures represented, the delegates want to acknowledge that the perspectives included in the conference represented only a small fraction of the world population and did not include perspectives from Eastern Europe, the Orient, the Third World, and several other Western nations, both European and American" ([15], p. 135).

Indeed, the delegates frequently disagreed on "the ethics" of various forms of treatment abatement. Thus the report concludes with a section on "Dissents." To be sure, the notion of *consensus* is used in that article's title in a rather "thin" sense as a shorthand for group unanimity. The author says: "All delegates rejected the simple vitalist assumption that prolonging life is always in a patient's interest" ([15], p. 135). It is this *unanimity* that serves as the expression of consensus – where every delegate agreed with some particular maxim or other at the time of adjournment. Consensus at the Appleton conference was understood as a *goal*, as an outcome sought, and not as a *process*.[1] Similarly, I shall maintain that consensus is not an end to be sought in (say) the deliberations of healthcare ethics committees (a view I share, in part, with Jonathan Moreno) [8], but rather that the process of consensus formation warrants our attention. This is especially the case when such committees

H.A.M.J. ten Have and H.-M. Sass (eds.), Consensus Formation in Healthcare Ethics, 35–44.

convene to review a case that involves the ethical predicaments of a patient in hospital. I simply caution that consensus formation in the context of policy development (another function of healthcare ethics committees) can easily lead to an unfruitful search for common agreement among the participants ([10], [11]).

From February, 1985 until June, 1987 the ethics committee at the Albert Einstein College of Medicine (Bronx, New York) struggled to resolve a question of policy that faced its institution: Whether pregnant Jehovah's Witnesses should be treated just like other adult patients with decisional capacity, or whether the existence of the fetus introduced a set of competing rights or interests ([7], p. 15).

In her detailed account of this deliberative process, Ruth Macklin, a philosopher at Albert Einstein and a member of the ethics committee, repeatedly refers to consensus, but in such a way that it stands as a synonym for agreement among the members of the ethics committee. Thus consensus reflected an *achievement* of the committee. Indeed, at a meeting of the ethics committee in July, 1987 the final version of the policy was passed, we are told, by consensus, i.e., common (unanimous?) agreement.

In the process of developing hospital policy this usage of "mutual agreement" is quite conventional, but neither philosophical nor very helpful. Consensus is surely other than mutual, unanimous, or majority agreement which "forms," or is reflected in, a vote of the majority present, or concretized by an expression of unanimity at the close of a committee's deliberations. Consensus is often a tacit process of committee deliberations as committees begin to work collectively.

However, before addressing this notion, we should be alert to the fact that the term 'consensus', even taken as a *goal* of deliberations, may refer to a variety of quite different "objects."

II. THE "OBJECTS OF CONSENSUS"

A deliberating healthcare ethics committee [14] may (1) arrive at universal agreement concerning the values to be applied in reviewing a particular patient's case or a particular hospital's policy; or (2) the committee members may discover that they all agree on how best to articulate the ethical values; or (3) reach virtual agreement with respect to the appropriateness of a particular medical treatment, (4) reach a unanimous view concerning whose best interests should be served in addition to the interests of the patient. After all, duties to the healthcare professionals in attendance may, according to the committee, *prima facie* override duties to the patient, notwithstanding the patient's stated preferences. Finally, (5) consensus can indicate that everything that need be said had been said during the discussion and that unanimous advice was offered; this is the "standard view" of consensus I mentioned earlier.

This notion of consensus formation can easily lead members to provide *recommendations* to or even to make decisions for the person acting on the patient's behalf, e.g., the physician or nurse who brought the ethical issue to the attention of the committee. Furthermore, there may be an additional objection to the notion of consensus as a goal or achievement: recommendations are proffered that go beyond the committee's legitimate *moral* authority. Let me explain.

On May 10, 1985 the Judicial Council of the American Medical Association went on record that healthcare ethics committees in the U.S. should be "advisory in purpose" and that they should only "consider and assist in resolving unusual, complicated ethical problems involving issues that affect the care and treatment of patients . . ." ([1], p. 2698). However, in its explication of this *advisory role*, the Judicial Council repeatedly used the term 'recommendations', which some may take as an even stronger mandate than a mandate "to advise only." Finally, and with qualification, the Council concludes: "The recommendations of the ethics committee should be offered precisely as recommendations imposing no obligation for acceptance on the part of the institution, its governing board, medical staff, attending physician, or other persons" ([1], p. 2669). A year later, in 1986, Professors Kliegman, Mahowald, and Youngner at Case Western University (Cleveland, Ohio) repeatedly referred to the activity of *recommending* by ethics committees; pausing to note that such recommendations are "advisory (rather than obligatory)" and "remain advisory not mandatory" ([6], p. 184). The authors offer a series of guidelines and end by suggesting that ethics committees summarize their case review sessions by articulating the consensus view, or the minority view (with actual mention of these views), or register the disparate views if neither consensus nor majority conclusions are reached.

I have been preparing to argue that to proffer recommendations is antithetical to the very *process* of case review that healthcare ethics committees pursue. By now it is obvious that for me consensus is not the proper goal of healthcare ethics committees' deliberations; if it were, we could simply cut most meetings short by asking everyone present to capitulate to a single standpoint so that, in the end, everyone agreed and no one dissented. Like Jonathan Moreno (who has seriously attended to this notion in *Deciding Together* [9] and other essays [8]), I think this is unsatisfactory. But unlike Moreno, I do not believe that ethics committees in healthcare institutions are typically "consensus driven." Therefore a call for the abandonment of the search for consensus as a goal is moot. But I do believe that committees are frequently too consensus driven, too goal oriented. Since this is an empirical matter I leave it to social scientist researchers. Moreno's view, then, can at the very least serve as an admonition; in this respect repeating his warning can do no harm.

III. THE RATIONAL SEARCH FOR COHERENCE

If the goal of group consensus is the only one of interest to healthcare ethics committees, they ignore the social dynamics at their peril. For a formal vote is an inappropriate gesture following the often arduous task of analyzing the medical and ethical aspects of a given patient's case; moreover, it is not obvious what constitutes the *process* of consensus formation that many believe drives a committee's deliberations. This obscurity may be due to committees' not seeking consensus formation after all; rather, they seek *coherence formation* (a notion I introduce here and now explicate).

Discussions of a particular patient's case frequently turn to the patient's life – his or her interpersonal life and lifestyle. Committee members typically ask questions that signal a search to *reconstruct* the essential elements of the patient's life – personal history, social relationships, medical history, psychological response to the present illness, transcendent concerns, especially concerns voiced when dying. Coherence formation is, then, the careful search for the integrating elements of a patient's life and lifestyle, his or her full biography, with special attention to the normative aspects of his life as *he* (if competent) determines them. Given a committee's commitment to uncover the life and values of the patient, the search or quest for coherence, any "quest for the 'right' decision" pales [4]. That is, a careful description of the actual deliberations of healthcare ethics committees, given their multidisciplinary composition, reveals a quest for *coherence*. It is this quest that makes vote-taking unseemly. Moreover, I reject the notion of majority rule (i.e., combining individual advice into a single recommendation) as objectionable here because, the outcome involves only the patient and a few others quite specifically and personally ([2], p. 236). Moreover, any method of combining decisions after they have been made on an individual basis is prudentially inadvisable. This would bypass the utilities and probabilities that initially went into separate contributions of the members. In short, the process I call "coherence formation" should, in the end, reflect the reasoned and compassionate responses of the representative members' variegated experience, expertise, and ethical reflection.

In 1984, Peter Singer and Deane Wells criticized the Ethics Advisory Board (U.S. Department of Health, Education and Welfare) in its attempt to reach consensus concerning the moral status of the human fetus. They described the Board's contradictory propositions as a reflection of "a masterpiece of consensus drafting" ([13], p. 198). Here the Australians understood 'consensus' as a goal, since they could readily show that the EAB could *not* agree on the moral status of the fetus. However, Singer and Wells pleaded for "more rigorous argument" while they recognized the fallibility of human rationality ([13], p. 199). In short, the sometimes unfeasible process of reconstructing the life and normative features of that life of the patient whose predicament is brought for discussion warrants the most competent expertise available. Thus "the only kind of consensus that is worth while," according to Singer and Wells, is consensus "built on respect for the role of reason and argument in reaching conclusions" ([12], p. 203). Consensus under this construal is clearly consensus as a process (an idea these thinkers proffered in 1984!).

If Singer and Wells are to be taken seriously, then it is necessary to mention the importance of criticism of the views of each member by the others. Because those who value a democratic polity value criticism and tolerance so highly, the notion of "reaching a consensus" becomes far less important and even far less interesting. Notwithstanding their intrinsic limitations, committees can frequently best serve as honest brokers among contending values, norms, ideas, and forceful assertions. As a result of this process, healthcare ethics committees serve a caring and nurturing function. As these committees work to assist in the resolution of moral disagreements and controversies, they can only succeed in an atmosphere of mutual respect and tolerance. Those who participate in such deliberation eventually appreciate working through the multifarious ideas, values, and suggestions directed to the ethical concern and care of patients. Consensus as a process, then, also implies *mediation* among those participants who frequently espouse a wide variety of moral commitments. Since the work of healthcare ethics committees occurs primarily in a pluralist secular milieu, the members must remain aware of their strictly interpersonal or *social authority*, their clear lack of moral authority, and remain vigilant that they not uncritically accept any legal authority – since many are eager and willing to bestow such legal authority upon them.

IV. SOCIAL AUTHORITY, NOT MORAL AUTHORITY

When the state of New Jersey's Supreme Court, in 1976, called for the establishment of a committee to determine the prognosis of Ms Karen Quinlan's illness, it created confusion about the appropriate role and authority of such "ethics" committees. Indeed, we continue to remain confused about the appropriate role of healthcare ethics committees [14] when they convene to discuss and review cases of patients who are caught at the epicenter of clinical, ethical and (sometimes) legal currents.

Given our present understanding of the function of case review and analysis, healthcare ethics committees have not to my knowledge indicated (in publications) that they have been confronting the *moral* status of consensus (though Jonathan Moreno has done so in his article, "Ethics by Committee: The Moral Authority of Consensus" [8] and in *Deciding Together* [9]). According to Moreno, "consensus should be thought of primarily as a condition of deliberation rather than as its goal" ([8], p. 412). In the closing pages of his essay, Moreno asserts that, "The point of ethical deliberation is not to reach consensus but to attain a desirable end, an end that settles a controversy without further disagreement." Moreover, he adds "along the way there must be agreement about the soundness of the method being used. Thus consensus is not an abstract end. It conditions a *process of cooperative reconstruction* of a troubling situation into one in which whatever latent values there are can be recognized and, by taking some action, perhaps more fully enjoyed" ([8], p. 428).

I have no quarrel with him; indeed, I applaud his employment of the Deweyan notion of consensus as a process of cooperative reconstruction. But it is important to underscore that what warrants this group effort is not some form of moral authority granted from "on high." So far, healthcare ethics committees have no formal *moral* authority through which to take decision and recommendations on patients or their representatives

[5]. These committees, therefore, are not warranted in making decisions for others; they should refrain from proffering recommendations that send this message, and from giving the impression that they possess moral authority. At most, such committees have *social authority*, and therefore they should function in a *strictly advisory* capacity. This is an important role.

Students of group process know that committees are "subject to multiple sources of correction" ([12], p. 5). Moreover, ethics committees are not tainted as a "group of experts." Hopefully they send the message "that the hospital community is one of ethical concern" ([12], p. 10). In short, the members of healthcare ethics committees, as Judith Wilson Ross put it, "are the creators [and I might add, the product] of the moral community" ([12], p. 15).

V. CONCLUSION: SOCIAL PRESSURE ON ETHICAL JUDGMENTS

In 1952, Solomon Asch published *Social Psychology* and woke his readers from their everyday social slumber. Quoting from the French sociologist, G. Tarde, Asch reminded us that "Social man is a somnambulist" ([3], p. 398). For our purposes this remark should serve to warn us that healthcare ethics committee members are akin to sleepwalkers. We may attend to the procedural and substantive issues as we participate in committee activities; we may consider the goal and the process of consensus or coherence formation; we may even succeed in remaining tolerant of the plurality of moral views with which we strongly disagree during our committee discussions; we may even pledge ourselves to remain always aware of the various ways our deliberations bear on the patient's best interests, or his or her family's interests, or even on the welfare of the hospital staff and the local community. But however worthy in themselves, these come to little if during actual committee discussions some of the members are made to capitulate to the judgments of others. This is especially true when these judgments are rationally unpersuasive. In short, we are far too susceptible to suggestion when we are attentive and engrossed in the voices around us. For, far too often we tacitly induce effects in others with the forceful expression of our assertions; far too often we assent to the beliefs of others in virtual uncritical acceptance. We are, in short, all quite capable of being indirectly affected by external impressions. Indeed, Asch demonstrated that the majority of the members of a group could quite easily influence the judgments of a minority of its members. By listening to the judgments of the majority, individuals may alter their own judgments, whether or not these views are the more illuminating, rational, or even central to the patient's ethical dilemma.

The results of Asch's experiments bear heavily on group processes in general; all the more profound are the implications of his work for those of us who are involved in ethical deliberations!

As space is unavailable here for me to dwell on the various factors that tend to influence the judgments (including the moral judgments) of healthcare ethics committee members – e.g., the status or prestige of physicians who attend these meetings, or whether there are other specific reasons for the influence of one subgroup on other committee members – suffice it to say that *trust* in the other members by each member him or herself is a necessary ingredient in this social process in order to assure that undue influence is not exerted under the guise of the best interests of the patient.

In *Social Psychology*, Asch introduced the terms 'independent' and 'yielder'. The goal of committees is for each member to relate to the others as independents, not as yielders, the latter being dependent or overly relying on the judgments of others. The everpresent danger that independents will coerce others to *yield* must also be kept in mind during these committee deliberations.

Returning, then, to the distinction between consensus as a goal of deliberation and consensus as a process that continues throughout deliberations, it should be clear by now that if consensus is not to be an empty and even a treacherous gesture, it must have moral validity, i.e., "the meaning of consensus collapses when individuals act like mirrors that reflect each other" ([3], p. 495). The integrity of every member is indispensable in seeking rational (if not always mutual) agreement; that is, when individuals have no way of reaching mutual agreement they should be free to disagree. That is, the act of yielding is antisocial because it leads to error and confusion. The path of independence, however, demands self-assertion especially when confronted with often intractable ethical dilemmas in healthcare. Yielding, on the other hand, involves renunciation of the self. Yet no matter how independent we may take ourselves to be – thinking only other members are yielders – we must always remain on guard against those social forces and pressures that would denigrate our ethical judgment and understanding for the sake of far lesser social ends. We can all be challenged here, even and especially if we take ourselves to be unflinching and "true" independents. Even the famous astronomer and scientist, Tycho Brahe, in 1572, did not fully take himself to be an independent, requiring the confirmation of his initial judgment by others, when he judged that he had made a new observation in the sky:

> Last year, in the month of November, on the eleventh day of the month, in the evening, after sunset, when, according to my habit, I was contemplating the stars in a clear sky, I noticed that a new and unusual star, surpassing the others in brilliancy, was shining almost directly above my head; and since I had, almost from boyhood, known all the stars of the heavens perfectly (there is no great difficulty in attaining that knowledge), it was quite evident to me that there had never before been any star in that place in the sky, even the smallest, to say nothing of a star so conspicuously bright as this. I was so astonished at this sight that I was not ashamed to doubt the trustworthiness of my own eyes. But when I observed that others, too, on having the place pointed out to them could see that there was really a star there, I had no further doubts ([3], p. 493).[2]

Center for Medical Ethics and Health Policy
Baylor College of Medicine
Houston, Texas, U.S.A.

NOTES

1 I am reminded of an essay by R. Morison ("Death: process or event?") and commentary by L.R. Kass ("Death as an event...,"), that appeared in *Science* (173, Aug. 20, 1971:694-98 and 698-702, respectively) in which the authors discuss whether death was a process or event. The outcome of the discussion revealed the complexity of the moral issues involved in the so-called "right to die" debate of the day. It was clear that the *process* of dying had been given short shrift in the extant literature on the topic: it was the arduous *process* of dying, not the moment of death, that required more serious attention.

2 Asch notes that he excerpted this quotation from H. Shapley and H.E. Howarth's. *A Sourcebook in Astronomy*, McGraw-Hill, New York, 1929.

BIBLIOGRAPHY

1. American Medical Association (Judicial Council): 1985, 'Guidelines for ethics committees in health care institutions', *Journal of the American Medical Association* **253** (18), 2698-2699.
2. Albert, D.A., Munson, R., and Resnik, M.D.: 1988, *Reasoning in Medicine: An Introduction to Clinical Inference* (The Johns Hopkins Series in Contemporary Medicine and Public Health), Johns Hopkins University Press, Baltimore, MD.
3. Asch, S.E.: 1952, *Social Psychology*, Prentice-Hall, Englewood Cliffs, NJ.
4. Ayres, S.: 'When values compete: Ethics committees and consensus', *Health Progress* **65** (11), 32-34.
5. Engelhardt, H.T., Jr.: 1996, *The Foundations of Bioethics*, 2nd ed., Oxford University Press, Oxford, UK/New York, NY (see especially pages 65-84).
6. Kliegman, R.M., Mahowald, M.B., and Youngner, S.J.: 1986, 'In our best interests: Experience and workings of an ethics review committee', *Journal of Pediatrics* **188**, 178-188.
7. Macklin, R.: 1988, 'The inner workings of an ethics committee: Latest battle over Jehovah's Witnesses', *Hastings Center Report* **18** (1), 15-20.
8. Moreno, J.D.: 1988, 'Ethics by committee: The moral authority of consensus', *The Journal of Medicine and Philosophy* **14** (4), 411-432.
9. Moreno, J.D.: 1995, *Deciding Together: Bioethics and Moral Consensus*, Oxford University Press, New York, NY/Oxford, UK.
10. Robertson, J.A.: 1984, 'Committee as decision makers: Alternative structures and responsibilities', in R.E. Cranford and A.E. Doudera (eds.), *Institutional Ethics Committees and Health Care Decision Making*, Health Administration Press, Ann Arbor, MI, pp. 85-95.
11. Robertson, J.A.: 1984, 'Ethics committees in hospitals: Alternative structures and responsibilities', *Quality Review Bulletin* **10** (1), 6-10; reprinted in *Connecticut Medicine* **48** (7), 441-444.
12. Ross, J.W.: 1990, 'Case consultation: The committee or the clinical consultant?', *HEC Forum* **2** (5), 289-298.
13. Singer, P. and Wells, D.: 1984, *The Reproduction Revolution: New Ways of Making Babies*, Oxford University Press, New York, NY/Oxford, UK (see pages 196-204).
14. Spicker, S.F. and Ross, J.W. (eds.): 1989-1998, *HEC (Healthcare Ethics Committee) Forum: An Interprofessional Journal on Healthcare Institutions' Ethical and Legal Issues*, Kluwer Academic Publishers, Inc., Dordrecht, The Netherlands, volumes 1-10.
15. Stanley, J.M. *et al.*: 1989, 'The Appleton consensus: Suggested international guidelines for decisions to forego medical treatment', *Journal of Medical Ethics* **15**, 129-136.

HENK A.M.J. TEN HAVE

CONSENSUS FORMATION AND HEALTHCARE POLICY

I. INTRODUCTION

The relation between policy and consensus is ambiguous. On the one hand, policy is needed because consensus in society concerning moral matters and regulating the interactions between citizens is deficient or even absent. On the other hand, policy will only be developed and successfully implemented as long as there is at least some consensus over some issues. Healthcare policy is no exception. It draws on basic values such as health, solidarity, and equity, in order to maintain or reform, re-direct or re-structure systems of care in society; but it can only work as long as there is at the same time sufficient consensus regarding the basic character of values; and it will only be motivated out of a certain level of discontent with these values and with the mechanisms through which they are expressed in systems and structures. The task of healthcare policy therefore is twofold: to proceed from whatever foci of consensus are available, and to construct new areas of consensus. In this chapter, I will discuss the interactions between healthcare policy and consensus formation in reference to the debate on resource allocation in the Netherlands.

II. AN UNSYSTEMATIC APPROACH

In most countries over the last decade, the debate over health care resource allocation has increased its intensity. Professional organisations, advisory health councils, political parties, governmental committees, public funding boards and medical associations have published reports on the rising costs of medical care together with policy suggestions on both the allocation and rationing levels. Almost every respectable institution in the health care area has articulated, clarified and advertised its opinions regarding the allocation of scarce resources. The result is a kaleidoscopic landscape with heterogenous points of view.

But what is most striking is the gap between theory and practice. On the one hand, various principles and criteria for allocating scarce resources are identified, classified, and analysed. On the other hand, ad hoc

H.A.M.J. ten Have and H.-M. Sass (eds.), Consensus Formation in Healthcare Ethics, 45–59

solutions, unsystematic and divergent policies are pervasive in health care practice, leading, for example, to different methods of budgetting, growing waiting lists, and diverse but mostly implicit criteria for patient selection.

Only recently have several attempts been made to develop a consistent and encompassing health policy focused on strategies for allocating resources and, in particular, on procedures for setting priorities. In retrospect, a phase of unsystematic variety of opinion and dissensus of ideas apparently is necessary to discover and find foci for building a new consensus in order to develop explicit policies. Such an unsystematic approach will not result in specific measures or regulations which can be implemented; but it will create an important condition for these practical policies, *viz.* agenda-setting. While the issue of resource allocation is at first a highly complex and intricate conglomerate of problems and questions, through a plethora of debates certain issues emerge as fundamental, pervasive or primal. The policy response to such issues is not yet clear, but it has become clearer what has priority. The identification of basic questions, and the growing consensus about what are the basic questions, therefore is the ground-work necessary for later practical policies.

The establishment and work of the Dutch Governmental Committee for Choices in Healthcare is an illustration of this new phase in the health care allocation debate. The Committee, installed in August 1990 and reporting in November 1991, was invited to develop strategies for making choices between existing and new medical possibilities in health care. Three questions were put on the Committee's agenda: 1) Why make choices in health care?; 2) Between what services do we have to choose?; 3) How should we make choices?

The strategies developed by the Committee in answering these questions should furthermore initiate a broad public debate about the relative necessities of services in the health care system. As an example of a priority setting procedure the Oregon Health Care Plan has been studied for possible application in the Netherlands [3].

III. IDENTIFYING THE BASIC QUESTIONS

The formulation of the Committee's task shows the pattern of basic issues which has emerged in the preceding unsystematic phase of the debate:

(a) policy should focus on choices in health care, more than on limits to health care;
(b) medical services can be categorised as more and less necessary or basic;
(c) emphasis should be on the priority setting process rather than on the product (i.e., a priority list).

However, all three questions submitted to the Committee presuppose that it is necessary to allocate resources or to make choices. Precisely, this presupposition is specifically addressed by the Committee as the most basic problem [6]. Before adequate policies can be developed, there should be broad public awareness that choices are unavoidable. Before the other questions can be answered, there should at least be some minimal consensus about the necessity of making choices and setting priorities. Crucial to the formation of this consensus is the interpretation of the allocation problem as more than simply an economic issue.

The necessity of allocating and choosing

Many are not convinced that choices in health care are unavoidable. Research data make clear that the majority of the Dutch population is opposed to making choices in health care: 55% agree that every treatment should be available regardless of its costs and regardless of the probability that it has a curative effect [9]. Moreover, 51% are prepared to pay double the current health insurance premium if that could guarantee the availability of every treatment possible. An even larger majority (78%) disagree with choosing amongst expensive medical technologies if this would make those technologies only affordable for higher income groups. A situation in which income will be a decisive factor for access to health services is strongly opposed. The major problem identified in health care is the shortage of personnel (and not the increasing costs of health care).

A similar questionnaire among health care professionals produces more or less the same results: for 50% of the respondents a further increase of health care costs is acceptable; 66% agree that people should pay more for health care [12]. A large majority agrees that access should not be related to income, but only 41% agree that every treatment available should be provided. Almost all physicians (92%) state that the quality of health care in the Netherlands is good or even excellent. But at the same time a majority acknowledges that there are too many treatments with low or marginal benefit (63%), that the use of diagnostic procedures is overrated (86%), that public expectations concerning medical technology are too high (82%). One of the major problems in health care identified by health professionals, is consumerism.

These data indicate that the crucial issue in the resource allocation debate today is the perception that making choices, at least in health care policy, is not really necessary. This perception is influenced by two specific ways of interpreting the problem of scarce resources.

The first construes the problem as *financial scarcity*. In this case, three different types of analysis can be distinguished.

(a) Policy-makers refer to economic constraints as a hard datum; financial resources are always limited; there are competing social goods and the opportunities to satisfy all desirable goods are restricted; making choices is therefore unavoidable.

(b) A second analysis shows the relativity of scarcity: scarce resources are not an objective reality leading to the necessity of choices, but they are themselves the result of implicit a priori choices; scarcity is a human construct resulting from deliberate human limitations and decisions.

(c) A third analysis depends on distinguishing real and fictitious scarcity. Scarce resources that are problematic now, are in fact the result of inefficient use of available resources; for the time being, scarcity is not a real problem; what is needed before we start making painful choices, is a large-scale operation of making the delivery of health care more efficient.

The second interpretation emphasizes that the issue of scarce resources is primarily a *cultural problem*. Allocative questions will not be less problematic if more resources, money and manpower are available. The problem will not disappear if every health service is delivered in the most efficient way possible. The reason is that the basic problem is not scarcity of resources but the unlimited nature of human needs. The issue of allocation of resources is in fact a symptom of the prevailing value-system of modern societies. This system is in part common to all western societies, partly specific for Western European countries. In particular four values are crucial and underscore massive public support:

(a) health

For modern man, health is apparently one of the most important values in human life. Value research in the Netherlands, for example, indicates that 57% of the population identifies health as the most important value in life (data from 1987; compared to 36% in 1966 and 49% in 1981) [8]. Such valuation of health is related to medicalisation and proto-professionalisation: many individual and social problems are rephrased and defined as medical problems, and many people can only present problems in the medical language of complaint, illness, and disability. Through the media, attention is focused on the latest medical developments, thereby increasing claims to medical assistance.

(b) right to healthcare

As a result of policies extending health care coverage to a progressive number of people during the years of economic expansion and growth, and as a result of certain changes in the social and moral arena, many have come to feel that they have a *right* to adequate health care and medical treatment, and that by virtue of this right the provision of services in this area is a task morally incumbent on the community and therefore on the various agencies of government. They feel that measures such as cutting costs by raising financial barriers to limit access, discontinuing support to expensive services and health programmes, or selecting or refusing patients for scarce medical treatment, constitute grave injustices to those individuals who, as a consequence of these measures may find themselves deprived of care and treatment.

(c) equal access

The medical profession as well as the general public in the Netherlands is opposed to many proposals for rationing policies, budget cuts and cost-containing measures. This opposition is mainly inspired by a concern for the maintenance of equal access. Access to health care in the Netherlands is not contingent on income or on geography. However, accessibility implies an ongoing critical assessment of every innovation and new service. Innovative medical treatment which has experimentally been received by some, will soon be demanded by the public to be made available to all. Accessibility also implies a continuous debate on the content of basic needs and the extent of a minimally adequate health care – a debate which is hampered more often than not by a second motivation to oppose rationing proposals: personal interests. To every individual patient his or her own needs always appear to be basic, and this seems to be particularly the case when the needs involved entail life prolonging treatment. As in other countries, in the Netherlands, patients with similar needs are apt to organize themselves into pressure groups and to raise a hue and cry when their needs are not provided for. Concern for the quality of care should be mentioned as a third motivation for opposition against rationing. Quality of care is threatened primarily in health care settings serving the chronically ill and disabled elderly and the mentally deficient.

Opposition to making choices in health care is therefore partly explained by the importance given to equal access for all; in contradistinction to health care in the U.S., equal access for everyone is common practise in many European countries. The fear that this fundamental principle may be threatened by the supposed need to make choices, could explain much resistance against allocative policy proposals.

(d) solidarity

A value that is foundational to the health care system in the Netherlands is solidarity, understood as a collective obligation to care [5]. This group-oriented responsibility is still generally endorsed in Dutch society. In the present health insurance system 'solidarity' refers to income and risk factors like employment, age, and health. Recent data show around 87% of the population acknowledgeing the view that neither health nor age should influence the amount of premiums to be paid for health insurance [7]. Around 70% subscribed to the view that higher income-groups should be made to pay higher premiums than lower income-groups, a view which also turned out to be endorsed by two thirds of all respondents from higher income-groups. The old type of financial solidarity is thus not really disputed. Opposition to allocative policies is partly motivated by the idea that solidarity may be undermined through the creation of a two-tier system of health care with the more expensive technologies being available for the elite only.

From limits to choices

It is perhaps useful to make a distinction between the limits *of* care and the limits *in* care, referring, respectively, to limiting the health care system which is regarded as a responsibility of the government, and to limiting care for individual patients or patient categories which is considered to be the responsibility of health care professionals. This distinction implies that *medical* ethics as a discipline primarily concerned with individual welfare can only function within a more encompassing framework of *healthcare* ethics which is primarily concerned with the general welfare [4].

It is often argued that intrinsic factors underlie rising health care costs: specialisation, professionalisation, medicalisation and above all, technological innovation. Within the present system such factors generate almost unlimited claims. Notions such as 'customary in the profession' and 'whenever a medical indication is prevalent' seem to be the only criteria available to differentiate between claims which are supposed to be justified and those which are not. On this basis a health care policy of equal access and financial solidarity produces an almost uncontrollable system.

One way suggested to regain control is by setting *external* limits: budgetting systems, review organisations, technology assessment. However, it is unclear whether this external approach will be successful as long as there is no evolution from external to *internal* limits. The argument is that we must learn autonomously whether or not to restrict our claims to health care, and whether or not to withdraw from the system of scarcity which is in large part maintained by an obsession with longevity in the sense of 'surviving others', living longer than, and thus outliving others as a consequence of an inability to integrate and accept death and suffering as a component of life. Since the concept of autonomy may be, and in many European countries is linked with the concept of solidarity, self-determination may be considered to imply an ability to restrain our claims for the benefit of others. In its Kantian interpretation 'autonomy' denotes the capacity of setting limits to one's own behaviour and one's own resolutions: my health may not necessarily be a prime value. Self-determination should involve responsibility for the self-realisation of other members of the community. It is necessary to develop a new ethos of critical use, moderation and temperance in health care.

The rhetoric of setting limits is therefore inadequate for several reasons:

- It wrongly assumes that medicine and health care are relatively autonomous activities with their own specific internal dynamics which can be influenced by rigorous external constraints;
- It is defensive, assuming that governments can only limit and control the growth of medicine but can hardly influence the development of new technologies and promote critical use of available services;
- It can be indiscriminating, focussing attention on restraint and control of the system as a whole, without distinguishing between what is more and what is less beneficial within the system itself.

Introducing the rhetoric of 'making choices' represents a more positive attitude towards the problem of scarce resources, introducing a voluntaristic and rational approach, emphasizing that growth of medical knowledge is the result from deliberate options chosen by individual scholars, policy makers and subsidizers, recognizing that governments have in fact encouraged some developments and discouraged others by regulations and financing policies, and inducing every actor to select positive opportunities from everything modern medicine makes possible.

The priority setting process rather than the product

Senate Bill 27, part of the 1989 Oregon Basic Health Services Act, required the establishment of a Health Services Commission. The task of the Commission was to provide "a list of health services ranked by priority, from the most important to the least important, representing the comparative benefits of each service to the entire population to be served" [11]. Driven by a feeling of urgency and sense of pragmatism, the HSC provided subsequently improved versions of a priority list.

In a European context, the Oregon health policy goes several steps too far. First, it assumes that there is broad public awareness that making choices is unavoidable, – and I have argued that the lack of such awareness is a central problem in many Western European countries. Second, it assumes that there is consensus on the moral desirability of approaching the issue of scarce resources by setting priorities, rather than by rationing, waiting lists, and patient selection. In many European countries it is too early to notice such consensus, although there is growing dissatisfaction with the current practice of rationing. In fact, this practice leaves the solution of distributive problems to the individual health care professional.

From a logical as well as moral point of view, it is desirable that allocative issues are approached from a health policy, *viz.* macro-level, perspective. This is the level of 'first-order determinations' that settle the scope of individual possibilities [1]. Prefering the macro-level is connected with its specific characteristics:

(a) Decisions have a bearing on patient categories, not directly on individual persons;
(b) Decisions require explicit criteria, equally applicable to and for everyone;
(c) Decisions are made within a public process of deliberation;

(d) As many actors and groups of actors as possible are involved in the decision-making process.

Because of these characteristics, decision-making on the macro-level of health policy will give *prima facie* better guarantees for equal treatment of individuals than on the micro-level where specific and idiosyncratic factors may determine the outcome of the individual doctor-patient relationship. The same is true for fairness of distribution since the allocative criteria and procedures are more open and controllable through public inspection than on the micro-level. Next, the macro rather than the micro-level requires the development of a procedure on the basis of democratic involvement of all actors. Finally, developing a priority setting procedure on the macro-level underlines that the ultimate responsibility for allocative decisions has been accepted by society. That will be a significant revision of the current practice in which individual health care professionals are unvoluntarily attempting to solve problematic situations that they have not individually created.

These insights into the moral qualities of macro-level decisionmaking as well as the practical example of the Oregon prioritisation process play an important role in creating a consensual basis for healthcare policy. Although there is no substantial consensus about the rules, methods and contents of priority setting, at least there is growing sense that this might be the preferable procedure with which consensus concerning allocative issues can be reached.

Over time consensus seems to grow about what might be the preferable procedure to bring about consensus concerning allocated scarce goods in healthcare. The formation of (formal) consensus regarding the appropriate procedure to reach (material) agreement on what allocative decisions to be made, has been enhanced by healthcare policy, but *via negativa*. Through failure, inadequacy and undesirable consequences, other policy strategies towards distribution of resources have gradually been eliminated.

IV. QUESTIONING BASIC VALUES

Significant for the present state of the allocation debate is also the increasing awareness that scarcity of resources is not just a financial, management, or organisational problem. It is also, and perhaps first of all, a socio-cultural problem. Both doctor and patient are participants in a cultural process, and actors with a fundamental post-modern mentality, overvaluing the contributions of medical science and technology to the pursuit of human happiness and wellbeing, and believing medicine's promise to eliminate human suffering and mortality. This more encompassing view directs us to the philosophical context of the allocation debate.

Rising health care costs should therefore give an impulse to examine critically the power and goals of medicine, not only because of its promises and its actual contributions to diminishing morbidity and suffering but also because of its interference with human dignity and its transgression of the boundary of meaningful life. The recently acquired medical power should be counterbalanced by a new medical ethics and a new awareness of a more critical use of medicine's technologies. Purely economic arguments do not suffice; there should be philosophical arguments for choosing those medical services which promote, or at least are consonant, with basic social values. Even when society is willing to provide unlimited resources to health care, choices will be inevitable, since medical developments – new treatments and diagnostics, opportunities for prediction and prevention, possibilities for designing offspring and managing mortality – will increasingly question the basic values of that particular society.

Developing a correct procedure for setting priorities will not evade the problem that basic values are at stake. Some level of consensus should be accomplished in order to apply the procedure. Priority setting implies making distinctions between more and less important health care services, between essential and non-essential care, between necessary and unnecessary treatment options. The proposal to make these distinctions (though flowing from moral considerations) is, at least from a European perspective, not consonant with traditional moral notions as equal access and solidarity; it is therefore important to explain how these notions are compatable with making distinctions between health services that are now in many countries freely available within current insurance systems.

The formal consensus regarding the procedure of approaching the allocation problem can lead to practical policy only if it is consonant with the prevailing value-system of society, *viz.*, if it is linked with the preexistent material consensus regarding basic values in healthcare. Therefore, after having argued that choices in health care are unavoidable and that the best way to make choices is through priority setting on the health policy level, the next step in a European context is to show that making choices is at least compatible with equal access and solidarity. The fit between formal and material consensus can also be constructed with positive arguments. In its report, the Dutch Committee for Choices in Health Care argues that the best way to safeguard the realisation of these basic values in future health care practices is through making choices between what is more important in health care and what is less.

In times of scarcity the notion of equal access is inadequate: it furnishes little or no guidance on which rationing policies should be applied and which health care settings they should be applied to. By making choices in a priority setting process, equal access for everyone can be guaranteed to every service or treatment that is regarded as important or essential. The same holds for solidarity. By asking solidarity for every health care service possible and every medical treatment available without any reference to their necessity and benefit, the notion of solidarity will be stretched beyond reasonable and affordable borders, and thus will be self-defeating. Making choices in health care can revitalize the concept of solidarity and endow it with new meaning. Health policy today is in many ways involved in attempts to shift the burden of care from the state to the individual. In doing so, a new type of solidarity might be promoted: solidarity not in the sense of an endorsement of redistributions of income, but in the sense of a disposition to accept responsibility for one's own life and one's own choices in life. In its latter sense solidarity may become a reason for self-exclusion from care as well as a reason for private initiative in organising and financing self-care in new social support systems. This latter concept of solidarity would imply that the autonomy of the individual consists in a recognition that one's own interests may be best served by promoting the common good.

The starting-point for the Committee's argument is the proposition that everyone who needs health care must be able to obtain it. However, equal access to health care should not be determined by demands but by needs. In order to have a just distribution of services, it is not important *that* all services are equally accessible, but crucial is *what* services are accessible. Not every health care service is equally relevant for maintaining or restoring health. Thus it is important to identify 'basic care', 'essential services' or 'core health services' that are focused on basic health-care needs in contradistinction to individual preferences, demands or wants. Relevant needs should be distinguished from all the things we can come to demand or want. In his theory of health-care needs, Daniels argues that needs are distinct in relation to their object, *viz.* health [2]. The concept of health is therefore the most appropriate standard for characterizing health care needs. This focus also illuminates that health enables persons to maintain a normal range of opportunities to realize their life plans in a given society. Since health care services in itself are not 'basic' or 'essential', the Committee prefers the expression 'necessary', because it implies a relationship between the particular kind of care or service with a particular goal ('necessary for what?').

The Committee defines health in general terms as the ability to function normally. However, 'normal function' can be approached from three different perspectives.

(1) The individual approach

Here, health is related to autonomy and self-determination. It is the 'balance between what a person wants and what a person can achieve' ([3], p. 51). Defined as such, health can vary according to various individuals; its content depends upon individual preferences. But then, no distinction is possible between basic needs and preferences; what is a basic need for one will not be for another. This approach therefore is not helpful in determining on a societal level what is necessary care that should be accessible to all. Even if through a democratic decisionmaking process (such as in Oregon) the largest common denominator or the smallest common multiplier of individual demands could be determined, we would lack criteria to identify necessary care ([3], p. 51).

(2) The medical-professional approach

In this approach it is the medical profession that defines health, *viz.* as the absence of disease. This approach is defended by Daniels. He interpretes health as "normal species-typical functioning"; disease is defined as "deviation from the natural functional organization of a typical member of a species" ([2], p. 28). Basic functions of the human species are survival and reproduction. Health care is more necessary as it prevents or removes dangers to life and enhances normal biological function. In this approach, necessary care could be distinguished according to the severity of illness; this was in fact proposed as a criterion by a Norwegian Committee in 1987 [10], although the criteria of severity are debatable. Nevertheless, this approach has a tendency to neglect the psychosocial functioning of individuals. It is also questionable whether normal species-typical functioning can be defined regardless of the social circumstances.

(3) The community-oriented approach

In this approach, preferred by the Committee, health is regarded as the ability of every member of the society to participate in social life. Health care is necessary "when it enables an individual to share, maintain and if possible to improve his/her life together with other members of the community" ([3], p. 54). 'Crucial' care is what the community thinks is necessary from the point of view of the patient. This approach is not utilitarian because what is considered to be in the interests of the community is dependent on its social values and norms. Every community exists because it presupposes a normative, deontological framework defining the meaning of its interests. In Dutch society at least three normative presuppositions define the communal perspective: a) the fundamental equality of persons (established in the Constitution), b) the fundamental need for protection of human life (endorsed in international conventions), and c) the principle of solidarity (expressed in the organization and structure of social systems, particularly the health care system).

Given this normative framework, it could be specified from the perspective of the community what should be regarded as necessary care in the Netherlands. The Committee distinguishes three categories of necessary care: 1) facilities which guarantee care for those members of the society who cannot care for themselves (e.g., nursing home care, psychogeriatrics, care for the mentally handicapped); 2) facilities aimed at maintaining or restoring the ability to participate in social activities when such ability is acutely endangered (e.g., emergency medical care, care for premature babies, prevention of infectious diseases, centres for acute psychiatric patients); 3) care depending on the extent and seriousness of the disease; priority among facilities in this group depends not only on need, but additional criteria decide whether a facility would be included in the basic package. From a community-oriented perspective, the first category is more important than the second or the third, and the second more than the third.

V. CONCLUSION. CONSENSUS FORMATION WITHOUT POLICY

The proposals of the Committee intend to start a broad public debate on health care services. The Dutch Department of Health has allocated a substantial budget (several million guilders) to increase the number of participants in this discussion. Indeed, especially among organizations of women, patients, handicapped and elderly many initiatives and activities have been started. However, the political debate so far has been disappointing. The Cabinet response (in June 1992) focused primarily on promoting appropriate care, giving a major role to health care professionals in defining standards of care and treatment protocols [13]. The present government (since 1994) seems to prefer a restrictive, piecemeal approach, identifying and proposing selected services to be excluded from the basic insurance package. Each time, the coherence between formal and material consensus is extremely fragile. For example, the government's proposal to exclude contraceptive medicines from the basic insurance package elicited emotional debates in terms of the same basic values the proposal intended to preserve in the longer run. In the political arena explicit choices in health care can no longer be avoided. Although with each proposal debate will start again and initiatives will be taken to ameliorate its consequences, both antagonists and protagonists agree that this piecemeal approach is morally better than the current situation, where the medical profession has been given the task to decide in daily practice how to cope with scarce resources. These sorts of decisions cannot really solve the problems, only ameliorate the consequences. Without explicit decisionmaking at the macro level, the basic principle of solidarity will further erode. What is needed is a policy which guarantees equal access to services providing for communally-agreed necessary care along with special protection of vulnerable groups within the community in order to maintain equality of result and opportunity.

Department of Ethics, Philosophy and History of Medicine,
Catholic University of Medicine,
Nijmegen, The Netherlands

BIBLIOGRAPHY

1. Calabresi, G. and Bobbitt, P.: 1978, *Tragic Choices*, Norton & Comp., New York.
2. Daniels, N.: 1985, *Just Health Care*, Cambridge University Press, Cambridge, Massachusetts.
3. Government Committee on Choices in Health Care: 1992, *Choices in Health Care*, Ministry of Welfare, Health and Cultural Affairs, Rijswijk.
4. ten Have, H.A.M.J.: 1988, 'Ethics and economics in health care: a medical philosopher's view', in G. Mooney and A. McGuire (eds.), *Medical Ethics and Economics in Health Care*, Oxford University Press, Oxford, pp. 23-39.

5. ten Have, H.A.M.J. and Keasberry, H.J.: 1992, 'Equity and solidarity; the context of health care in the Netherlands', *The Journal of Medicine and Philosophy* **17**, 463-477.
6. ten Have, H.A.M.J.: 1993, 'Choosing core health services in the Netherlands', *Health Care Analysis* **1**, 43-47.
7. Janssen, R., van de Berg, J. and Haveman, H.: 1987, 'Solidariteit en het ziektekostenverzekeringsstelsel', *Gezondheid & Samenleving* **8**, 2-9.
8. Mootz, M.: 1991, 'Culturele determinanten van medische consumptie. Een verkenning van mogelijke ontwikkelingen', in Commissie Keuzen in de Zorg, *Kiezen en Delen*, deel 1, DOP, Den Haag, pp. 267-279.
9. NSS/Marktonderzoek: 1991, 'Keuzen in de zorg – een opinie-onderzoek', in Commissie Keuzen in de Zorg, *Kiezen en Delen*, deel 1, DOP, Den Haag, pp. 305-307.
10. Royal Norwegian Ministry of Health and Social Affairs: 1990, *Health Plan 2000*, Oslo.
11. *Senate Bill 27, Section 4a (3)*: 1989, 65th Oregon Legislative Assembly.
12. Tijmstra, Tj., Busch, M.C.M. and Scaf-Klomp, W.: 1991, 'Keuzen in de zorg: Meningen van beroepsbeoefenaars', in Commissie Keuzen in de Zorg, *Kiezen en Delen*, deel 1, DOP, Den Haag, pp. 309-316.
13. Tweede Kamer der Staten-Generaal: 1992, *Modernisering zorgsector. Weloverwogen verder*, Sdu Uitgeverij, Den Haag, 1992, 22393, nr. 20.

PART II

CULTURES AND CONSENSUS FORMATION

HENRIK R. WULFF

CONTEMPORARY TRENDS IN HEALTHCARE ETHICS

I. INTRODUCTION

> The most striking feature of contemporary moral utterance is that so much of it is used to express disagreements; and the most striking feature of the debates in which these disagreements are expressed is their interminable character. I do not mean by this just that such debates go on and on and on – although they do – but also that they apparently can find no terminus. There seems to be no rational way of securing moral consensus in our culture ([4], p. 6).

This is a quote from the beginning of Alasdair MacIntyre's book *After Virtue* from 1981, and it describes well the ethical debate as it "goes on and on and on" in most Western countries today. It raises the question which I shall consider in this chapter: Which are the necessary conditions for discussing ethical problems in a rational way, and how can we ever hope to reach some degree of consensus when we are faced with an ethical dilemma?

This contribution deals with healthcare ethics, but the problems which concern us as healthcare professionals cannot be viewed in isolation. The solution of complex ethical problems posed by modern medicine requires that we consider the basic ethical norms of our particular society and that we apply these norms to the problems at hand. Medical ethics is not an international discipline, although some members of the medical profession seem to believe that it is. We must take into account the differences which exist between different cultures.

II. PRECONDITIONS FOR A RATIONAL ETHICAL DEBATE

The words *ethics* and *morals* originally meant exactly the same thing, customs or manners, and in those days when Greek and Latin were living languages there was little need for endless moral debate. The citizen of the Greek *polis*, for instance, was born to play a certain role in society, and the ethos, which was fixed by tradition, was something to be de-

H.A.M.J. ten Have and H.-M. Sass (eds.), Consensus Formation in Healthcare Ethics, 63–72

scribed and illustrated – often in a dramatic form – rather than something to be discussed. Today we find a similar situation in some surviving so-called primitive societies where generation after generation live their lives according to a well-established pattern, and in such cultures there is no need for concepts like ethics and morals in their modern sense: ethics is social practice and social practice is fixed. This is one extreme where there is little scope for ethical debate, but there is another extreme which also tends to eliminate the need for ethical reasoning.

This opposite extreme is represented by the radically libertarian society. The members of such a society may show tolerance towards each other, but they recognise only one inviolable ethical rule and that is respect for the right to self-determination. In such a culture ethical debates are equally futile, except those which concern one particular topic: whether or not the respect for self-determination is violated.

Both extremes are, of course, theoretical. There never was and never will be a society where human behaviour is completely codified, and there can be no society where everything is tolerated apart from the violation of individual rights.

Fig. 1. Relationship between the scope for rational ethical discourse and the extent to which human conduct in a particular society is guided by a common moral tradition or ethos.

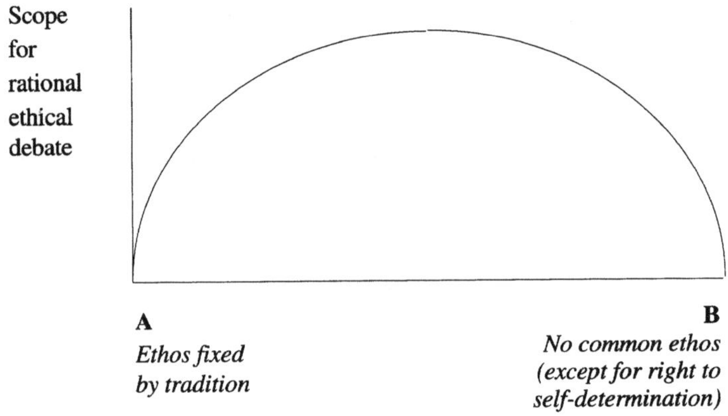

However, it may still be useful to imagine these extremes, A and B, as the ends of a scale (Fig. 1) and to consider the relative position of different Western societies on that scale. Orthodox Jewish communities with very fixed codes of behaviour are certainly closer to A than most Western societies, and even in my part of the world (Scandinavia) we had much stricter norms just a few generations ago. Then, people had to accept a fairly strict set of norms, and those who tried to break the rules were not expected to engage themselves in a rational dialogue, but were simply told "that it is not done" to behave in that manner.

There can be no doubt that most European societies have moved away from A towards B, and, possibly, we passed the highest point which gave the largest scope for rational ethical dialogues when our societies became thoroughly secularised and we started our search for moral standards not based on religious beliefs. The utilitarians appealed to happiness and desires. But what is happiness and what do we desire? Kant appealed to reason. But nobody can solve a complex dilemma in healthcare ethics by reason alone.

The society which has moved furthest towards B is that of the United States, and this development has had a strong impact internationally on medical ethics. Members of the medical profession all over the world are accustomed to reading American medical literature, and, since the American contributions to medical ethics are varied and often of a high quality, it is not surprising that they have had a profound influence on medical thinking in Europe. One sometimes gets the impression that European doctors with a particular interest in medical ethics have adopted American norms which are out of touch with those of the populations which they serve.

III. ETHICS AS INDIVIDUAL RIGHTS AND AS MUTUAL OBLIGATIONS

It was claimed above that the mainstream of ethical thinking in the United States is positioned closer to B on Fig. 1 than that of most European countries, and in this section I shall try to analyse the difference between the European and American tradition. This analysis, however, must be taken *cum grano salis*: that which will be called "the American way of thinking" originated in Europe where it is now being re-imported, and the contrasting "European way of thinking" has adherents among prominent American moral philosophers (e.g. Rawls [5]). Further, these "variants" of contemporary Western moral philosophy are closely related in so far as they are both based on the idea that man is an autonomous being, but the ethical implications of that fundamental idea are interpreted differently.

The traditional *American way of thinking* is characterised by four principles ([1,6]), which are usually listed in the following order suggesting a sequence of priority.

First, there is the *principle of autonomy*, which among doctors, if not among thoughtful philosophers, is simply interpreted as the right to self-determination. Then follows the *principle of non-maleficence*, which means that we must not deliberately harm other people. This principle, which is absolute, serves as a constraint to the principle of autonomy which otherwise constitutes the supreme principle. Next, there is the *principle of beneficence* which is not absolute. In one ethical declaration it is explained as follows: "All persons have some moral obligation to benefit others to some degree, including, perhaps even especially, those in need" ([6], p. 130). Finally, there is the *principle of justice*, which receives the lowest priority.

These four principles are sometimes called the *mantra* of medical ethics, but in fact they convey a political message; they convey the picture of a freedom-loving and possibly tolerant society, where people respect each other, but where social inequalities and social injustice cannot be prevented due to the relative and subordinate status of the two last principles. (If the principle of autonomy is substituted by the principle of sovereignty, the four principles also illustrate well the present state of international relations: The supreme principle is the sovereignty of each nation; war is forbidden; the rich countries of the first world have some duty to offer some assistance to the third world, but this duty is relative and the inequalities among nations remain).

This way of thinking has ancient roots, e.g. the philosophy of John Locke, and it is based on the idea of the natural rights of man. Expressed somewhat simplistically the argument goes as follows: man is autonomous, i.e. man has a free will; consequently, man has a right to do as he pleases, as long as he does not harm other people. This idea has great immediate appeal, but philosophically it is far from clear. We talk about the natural rights of man, but how does nature provide rights?

The other school of thought, which I have labeled the *traditional European way of thinking* is very different. It is based on the philosophy of Immanuel Kant, which stresses the duties rather than the rights of man. In other words, the Kantian argument runs somewhat like this: man is autonomous, i.e. man has a free will; consequently man has a *duty* to behave in a certain way. It is beyond the scope of this presentation to discuss this duty – the categorical imperative – in any detail, but I believe that it is fair to say that it is based on the idea of reciprocity and on the idea that each individual must take great care that he does not see himself in a privileged position. According to one of the formulations of the categorical imperative we must make the thought experiment that the moral principles which we adopt become universal laws; we must consider the implications of everybody acting in that particular way, including the possibility that others treat us as we treat them.

Kant regarded the categorical imperative as a product of reason, but, from a practical point of view, his message differs little from the *Golden Rule* of Christian ethics, as it was formulated in the Sermon on the Mount: Treat others as you would like them to treat you. According to this tradition, the four ordered principles mentioned above are, of course, totally inadequate. Justice, which before was at the bottom of the list, has moved to the top, as the just society may simply be regarded as a society where everybody treats each other according to this rule. The American philosopher John Rawls derives his concept of distributive justice from Kantian ethics [5], and his contribution to moral philosophy is yet another proof that the European and American traditions intermingle. Individual human rights – which are also valued by the adherents of this school of thought – are no longer derived in a mysterious fashion from the nature of man, but are seen as part of the reciprocal relationship between people living together in the same society.

IV. PRACTICAL IMPLICATIONS

Perhaps I distinguish too sharply between the two traditions, but it remains a fact that the organisation of the health service in the United States differs strongly from that of most European countries, and this difference is well explained by the philosophical distinction described above.

In the United States there is no national health service like the ones in Scandinavia and Great Britain, and their cannot be, as the establishment of these services is based on the idea of the high priority of distributive justice. There is, of course, a public health service in the United States, but it has a very different status. At the consensus conference at Appleton [6], for instance, the role of the public health service was described as follows: *it must provide an acceptable decent minimum of basic healthcare.* Such a health service is based on the non-absolute principle of beneficence, i.e. on a principle of charity, and to most Europeans (and to many Americans) it seems highly unjust. In this connection, it is worth noting that the proceedings of the Appleton conference, which also quoted the four principles mentioned above, were published under the heading "international guidelines", as if it was taken for granted that this way of thinking ought to be accepted universally.

The differences between the European and the American tradition may also be viewed from a different angle. If, for instance, somebody wrote a book about family ethics, then I hope that the author would stress the importance of the duties of the family members towards each others, and the presentation would be very incomplete, if it did not take into account such concepts as love and trust. Of course, children, wives and husbands also have individual rights, but the discussion of these rights would only come into the foreground when the relationship between the different members of the family had already deteriorated. Much the same can be said of the patient-doctor relationship, which primarily is based on mutual trust and the doctor's concern for his patient as a fellow human being. Of course, patients, patients' relatives, nurses and doctors also have rights, but discussions of these rights usually indicate that the primary ethical concerns have somehow been neglected. I certainly do not wish to imply that the patient-doctor relationship is less satisfactory in the United States than in Europe (apart from the fact that we do not have the uneasy feeling that a greedy lawyer is looking at us over the patient's shoulder), but it remains a fact that the presentation of most case histories in American literature on medical ethics treat the ethical problems in a quasi-legalistic manner. Typically, a case is presented where somehow the personal relationship between doctor and patient broke down, subsequently it is discussed at length who has the right to decide what, and finally the recommendation of the hospital ethics committee or the verdict of a court of law is presented as the solution to the ethical dilemma. No attempt is made to distinguish sharply between the ethical and the legal issues, and the real ethical problems – the factors which led to the breakdown of the mutual patient-doctor relationship – are rarely mentioned.

As stated already, I strongly believe that medical ethics is culture-specific and it is not my purpose to criticize contemporary American thinking which reflects a specific historical tradition, but I wish to warn European doctors and others responsible for our health services against the unwitting acceptance of transatlantic norms. We see this tendency today in Denmark, and I am sure that others have the same experience.

We have, for instance, recently had a lively debate about transplantation ethics, and it was interesting to see that the main concern was the possible violation of the rights of the donor. Considering the nature of the topic, one might have thought that the idea of reciprocity – of mutual obligations – would have been a better starting point for the debate. One might have linked the right to receive an organ (in case of need) to the duty to provide an organ (in case of brain death), but that did not happen.

We have also had a debate about research ethics which almost exclusively concerned the rights of the patients who take part in, for instance, randomised trials. The patient's right to be fully informed and to refuse participation must be respected – otherwise the doctor would violate the Kantian principle not to use other people *as a means only* – but the discussion is distorted when it does not also consider our duties towards each other. When we fall ill we benefit from the results of scientific studies done on previous patients, and therefore it must also be discussed to which extent we have some *prima facie* duty towards future patients.

V. THE FOURFOLD TABLE OF HEALTHCARE ETHICS

Table 1 serves to illustrate further the difference between the two traditions. Healthcare professionals have to reason both on the individual level and on the level of society, and they have to take into account both teleological and deontological considerations.

Table 1. The fourfold table of healthcare ethics

	Deontological reasoning	Teleological reasoning
Individual level	*Principle of autonomy*	*Seek best consequences for your patient*
Level of society	*Principle of justice*	*Seek best consequences for everybody*

Ethical reasoning in daily clinical practice mostly takes place on the individual level. Physicians are in each individual case professionally bound to seek that action which has the best consequences for their patient, but they also act under a deontological constraint: they must respect the patient as an autonomous person.

The medical or non-medical administrator, on the other hand, reasons on the level of society. He seeks the best consequences for everybody, which means that he is reasoning according to the principles of classical universal utilitarianism, but he also has to accept deontological constraints, in this case the principle of distributive justice or fairness.

I am aware of the fact that especially utilitarian moral philosophers will disagree with this presentation of the problems of medical ethics, as they will argue that considerations described as deontological represent teleological considerations on a higher level, but from a practical point of view the fourfold table illustrates well the quandaries of healthcare ethics. The physician may well feel quite convinced in the individual case that a certain action will have the best consequences for the patient, but he may still have to refrain from that action, if the patient disagrees. Similarly, the health economist may wish to maximize the utility of his actions on the level of society, but he also has to take into account, especially if he is employed in a national health service, whether he is ensuring a fair distribution of the limited resources.

The American way of thinking does to some extent offer a radical solution to the tensions that are illustrated by this table, as the principle of autonomy is interpreted as the principle of the right to self-determination. Then it follows, according to the four principles, that reasoning on the individual level must always take priority over reasoning on the societal level and that – on the individual level – the deontological constraint always has a higher priority than teleological considerations. If the right to self-determination is violated, the action is deemed paternalistic, and within this school of thought paternalism is almost used as an invective. Consequently, there is little scope for ethical debate.

In contrast, the Kantian or traditional Christian approach, which I believe still pervades much European thinking, offers no clearcut solutions. The right to self-determination is regarded as an important principle, but it is recognised that the principle of autonomy also has other facets. It may, for instance, sometimes be argued that the patient's future quality of life as an autonomous person is best served by a modicum of paternalistic action. It is realised that paternalism rooted in true compassion cannot always be rejected as unethical and that a human relationship based exclusively on the respect of rights is as cold as ice. It is also realised that ethics on the most fundamental level concerns the just relationship between individuals, for which reason the ethical considerations on the level of society cannot be ignored. The tension between "the rows" as well as the tension between "the columns" in the fourfold table remain, and it is these tensions which constitute the substrate of the ethical debate.

VI. CONSENSUS: THE TERMINUS OF RATIONAL ETHICAL DISCOURSE

At the beginning of this paper I raised the question: How can we ever hope to reach some degree of consensus when we are faced with an ethical dilemma? Nowadays, we need only consider the right half of the diagram (Fig. 1), and here the essential precondition for rational debate is some agreement as regards fundamental norms and values. Rational ethical dialogues involve deontological considerations which presuppose some agreement as regards basic duties and rights and they involve teleological arguments which presuppose some agreement as regards what is good and what is not good. If the participants in the dialogue share a set of fundamental norms and values, they may hope to reach a consensus by considering the facts, by analysing the problem and by balancing the different ethical considerations. They may, to put it briefly, succeed in reaching the same reflective equilibrium [3]. If they do not share a set of fundamental moral beliefs, their dialogue will, as Alasdair MacIntyre put it, just go on and on and on.

At present we see a "shift to the right" of the European moral tradition and, consequently, the scope for rational ethical discourse also seems to be waning in this part of the world, but the long-term prospects are far from clear. In the United States one notices a reaction in the form of calls for greater justice within the healthcare system (e.g. [2]), and generally speaking it is difficult to rid oneself of the suspicion that the very idea of respect for individual self-determination as the supreme ethical principle may be an expression of human *hubris*. Human beings, like other social animals, depend both on each other and on their environment, and a realistic appraisal of the present predicament of mankind may yet force us to emphasize more strongly the acceptance of mutual obligations for the common good. Such a development will once again enlarge the scope for rational ethical debate and it may even provide the terminus for such debates which Alistair MacIntyre sought in vain.

Department of Medical Philosophy and Clinical Theory
Panum Institute
University of Copenhagen
Copenhagen, Denmark

BIBLIOGRAPHY

1. Beauchamp, T.L. and Childress, J.F.: 1979, *Principles of Biomedical Ethics*, Oxford University Press, New York.
2. Callahan, D.: 1990, *What Kind of Life. The Limits of Medical Progress*, Simon and Schuster, New York.
3. Daniels, N.: 1980, 'Reflective Equilibrium and Archimedian Points', *Canadian Journal of Philosophy* **10**, 83-103.
4. MacIntyre, A.: 1981, *After Virtue. A Study in Moral Theory*, Duckworth, London.
5. Rawls, J.: 1971, *A Theory of Justice*, Oxford University Press, Oxford.
6. Stanley, J.M., *et al.*: 1989, 'The Appleton Consensus: Suggested International Guidelines for Decisions to Forego Medical Treatment', *Journal of Medical Ethics* **15**, 129-136.

HUB ZWART

MORAL DELIBERATION AND MORAL WARFARE: CONSENSUS FORMATION IN A PLURALISTIC SOCIETY

I. INTRODUCTION

In bioethical debate one often encounters the claim that contemporary society harbours many moralities, values and ideals. Therefore, it is considered pluralistic, and bioethics is considered an endeavour that strives for ethical consensus in a pluralistic society. Indeed, "consensus" and "pluralism" are prominent keywords in contemporary ethical and bioethical discourse. This chapter is devoted to the problem of consensus formation in a pluralistic society, while special attention is devoted to consensus formation in health care ethics (or bioethics). Before turning to the problem of consensus formation as such, I will briefly indicate how I understand the key terms "pluralism" and "consensus".

In this chapter, "pluralism" will indicate that we are faced with a situation where no single moral perspective is acceptable to all participants in moral deliberation. This conception of pluralism is in agreement with the claim that, in contemporary society, many individuals have become "moral strangers" to one another. It implies that, in ethical debate, it is no longer possible to make an appeal to basic and substantial moral insights, intuitions and experiences shared by all. Furthermore, I will stress the elaboration of pluralism by means of the language metaphor: fundamental moral conflicts are conceived as a collision, not merely between particular arguments, but rather between rival moral "languages", "idioms" or "vocabularies".

Closely tied to the concept of pluralism is the concept of "consensus". It refers to an instance of agreement, reached in a discursive way during a process of moral deliberation among proponents of differing views. To clarify this definition, two aspects must be emphasized. First, consensus is a kind of agreement that is beyond a mere compromise, as it implies a positive commitment on the part of those involved. Furthermore, it is an agreement reached in a pluralistic context, that is: in the absence of a single privileged or central moral viewpoint – in the absence of moral authority. In this respect it is interesting to note that "consensus" was often used during the Reformation period to indicate agreement reached

H.A.M.J. ten Have and H.-M. Sass (eds.), Consensus Formation in Healthcare Ethics, 73–91

among different branches of the Protestant movement, such as the *Consensus Tigurinus* (Zürich, 1549), the *Consensus pastorum Genevensium* (Genève, 1551) and the *Consensus Sendomiriensis* (Sandomir, 1570). Should there be a central viewpoint endorsed by all (which was more or less the case during the Middle Ages) those involved would probably prefer to talk about "truth" or "doctrine" instead of "consensus". The Protestant habit of identifying a particular instance of consensus by linking it to the locality where it was reached, recurs in contemporary bioethical discourse, where we encounter for instance the *Appleton consensus*, an agreement reached among bioethicist in the town of Appleton, Wisconsin ([11]; [12]).

It is the aim of this contribution to review differing views on consensus formation in a pluralistic society, stemming from differing conceptions of ethics and differing attitudes towards pluralism. Two views on consensus formation in health care ethics will be identified: the minimalist and the substantialist view. The import of both views will be further elaborated by comparing them to perspectives on consensus and pluralism in contemporary moral philosophy.

II. THE MINIMALIST VIEW ON CONSENSUS FORMATION

What are the prospects for consensus formation in a pluralistic society? One view on consensus formation is quite prominent in health care ethics. It is sometimes referred to as the "liberal" view. However, as the term "liberalism" contains some confusing connotations, I will refrain from using it. Other available terms to indicate this view are "procedural", "discursive", "minimalist", and "deliberative" ethics. Sometimes these terms are used as synonyms, on other occasions they emphasize different aspects of the view I have in mind. In this section I will try to identify the conceptual *nucleus* or quintessence of the view on consensus formation indicated by these terms. From the terms just mentioned I will, more or less arbitrarily, choose the term "minimalist ethics" to indicate this nucleus. I must emphasize however that "minimalism" must be taken in a descriptive, not in any pejorative sense. It is my contention that, although at first glance contemporary ethics might give the impression of being "ethics after Babel", the basic claims of minimalism are shared by a fair majority of participants in ethical discourse on health care issues. Nevertheless, the minimalist view on consensus formation is criticized as well, namely by those participants in the debate who opt for a broader conception of ethics.

According to minimalist ethics, "pluralism" indicates that, in contemporary society, several moral communities, ideals and vocabularies coexist – that is, it endorses the conception of pluralism marked out above. Each moral community elaborates a moral perspective and moral idiom of its own. To avoid moral conflict among these communities, the first commandment of a pluralistic society calls for tolerance and mutual respect.

The absence of a single moral viewpoint, acceptable to all, implies that pluralism poses a serious threat to society. In case of conflict, the peaceful coexistence of moral communities might give way to a condition of moral warfare – a moral Lebanon or Bosnia so to speak – were "force and fraud" become prominent virtues. How is this condition to be avoided? Contemporary ethics does not advocate the *enforcement* of a moral "Leviathan", but rather devotes itself to bringing about a moral climate of tolerance and peaceful cooperation. To be able to sustain such a climate, minimalist ethics claims that, even in a pluralistic society, there are some basic moral demands which can and should be accepted by every individual. These demands are conceived as "procedural" rather than "substantial". They are derived from the idea that moral warfare can be avoided by moral deliberation. The two most important demands, articulated by minimalist ethics, derived from the idea of moral deliberation and considered acceptable and imperative to all, amount to the following:

(1) *Willingness to participate in moral deliberation.* In a pluralistic society, individuals can never be forced to accept substantial moral views. However, we *do* have the right to expect that, in case of moral conflict, those involved will be prepared to engage in peaceful negotiations to settle the moral dispute. To be reasonable and accountable means to be willing to participate in an encounter with "moral strangers", consisting of the unrestrained exchange of arguments. In principle, of course, one has the right to withdraw from deliberation in order to live a private life or seek personal salvation in silence, but in real life such an option will be proof difficult to hold on to. In almost everything we do, others will somehow be involved.

(2) *Mutual respect among participants.* Deliberation among "moral strangers" is pointless, unless respect for each others' views is granted. Even if we can not understand the views of our adversaries,

let alone subscribe to them, we should grant them a fair chance in the debate.

These two moral demands are at the heart of the minimalist proposal to cope with the threat of moral warfare, posed by pluralism. They raise however an important question: what will be the agenda of moral debate? From what has been said it is clear that minimalism will try to evade the "broad", substantial issues. In the case of health care ethics, for example, a protagonist of minimalism would be reluctant to discuss the place and meaning of health and medicine in human life and in a good society. These issues are often explicated within a religious or otherwise "substantial" framework, implying some "metaphysical" claims regarding the true nature or moral condition of man. Minimalism claims that, in a pluralistic society, consensus on these issues is beyond our reach. The agenda of moral deliberation should be neutral towards religious and metaphysical commitments. Ethics should devote itself to the conceptional elaboration of the conditions for a fair exchange of arguments among moral strangers.

This answer, however, does not solve all the problems. In terms of the language metaphor another question could be raised: What moral language will be spoken in a situation of unrestrained deliberation? Moral languages actually spoken and language games actually played will convey implicit, yet substantial moral or even metaphysical views. If a moral language that might be considered neutral towards substantive moral claims and commitments cannot be found, minimalist ethics will fail to reach its goal, since it will reinforce the substantial views embedded in the dominant language. Therefore, the minimalist endeavour implies the quest for a neutral and common moral language that will be congenial and acceptable to all, regardless of the "broad" vocabularies we share with those to whom we feel morally akin. The neutral language will sustain pluralism and make it possible, rather than silence it. It will maintain and not endanger the peaceful coexistence among incommensurable moral views and idioms. It will not interfere with the religious or metaphysical vocabularies, provided they do not go beyond their proper limits (that is, provided they do not impose themselves on those who do not consent to the substantial views they convey). But perhaps this so-called neutral language will only *seem* neutral, merely because it happens to be the dominant one?

Indeed, the moral demands of minimalism articulated above, might seem trivial at first glance. Yet they contain some important implications for moral deliberation. One of the implications, already mentioned, is the endeavour to bring about and elaborate a neutral moral language that will render the peaceful coexistence of moral communities and incommensurable moralities possible, – a task which basically comes down to formulating and elaborating a discursive regime consisting of keywords and principles such as "autonomy", "tolerance", "respect" and "consent". A second important implication is that, in the minimalist view, ethics should not devote itself to the quest for virtues, ideals of life and other items of morality in a broader sense. These broader issues and concerns – "What is the good?", "What is a virtuous life?" – are "de-listed" [3] from the agenda of contemporary ethics and transferred to the private sphere, to the voluntary commitments of autonomous individuals towards particular moral communities.

These considerations can be articulated in a third moral demand of minimalism, which is clearly less trivial and more controversial that the other two. It amounts to the following:

(3) Participants in moral deliberation with moral strangers are expected to *refrain from any appeal to a particular moral sense or to particular moral intuitions, or to basic and substantial moral insights, or to claims derived from broad conceptions of life*, unless they can be sustained or elaborated by means of reason and argument, because such intuitions, insights and claims will not prove convincing to "moral strangers" who do not share the basic, substantial convictions that sustain them.

This is a questionable demand, because we are expected to refrain from an appeal to the very considerations which, in our personal moral decisions, are often the decisive ones. Although in a pluralistic society one is allowed to articulate his or her personal convictions and commitments, in a situation of moral deliberation among "moral strangers" they are not expected to contribute to consensus-formation.

It is this third claim of the minimalist view, the one concerning the idiom and the agenda of deliberation, that is the most severely contested and disputed one. A "substantial" view on consensus formation will eagerly endorse both the willingness to participate in deliberation and the demand of mutual respect, but will never endorse such a restrained idiom or agenda, never accept these "terms" of the debate. The substantial view will claim that moral arguments are never "neutral" but always imbedded in broader frames of reference, and that even the idea of moral deliberation itself is not neutral but culturally and historically embedded. It will claim that we do not respect a participant in moral deliberation if we ask him to refrain from those considerations he himself considers pivotal, merely because we consider them too "broad" to contribute to consensus formation.

I will draw attention to one particular version of the minimalist endeavour, elaborated in bioethics by H.T. Engelhardt and challenged by another bioethicist, Daniel Callahan. Next, I will turn to contemporary moral philosophy to emphasize the significance of their dispute. I will briefly refer to the work of German (Apel and Habermas), French (Lyotard) and American (MacIntyre and Stout) philosophers.

III. CONSENSUS FORMATION IN HEALTH CARE ETHICS: THE ENGELHARDT/CALLAHAN DISPUTE

The minimalist view on pluralism seems rather dominant in contemporary bioethics. H.T. Engelhardt is a prominent spokesman of this view [4]. Modern society, he claims, can no longer hope to develop a single moral viewpoint that will prove convincing and acceptable to every individual. Therefore, bioethics should provide us with a *neutral common language* that might serve as an idiom for moral deliberation among advocates of incommensurable views. Such a language enables them to argue, in spite of their differences. It renders the peaceful coexistence of divergent moral communities possible. It serves as a vehicle for peaceful negotiation among protagonists of incommensurable moral views.

It is important to note that Engelhardt draws a distinction between society at large (the pluralistic society) and the many moral communities *within* this society. Every moral community elaborates a particular moral perspective of its own. This particular moral perspective seems binding and inescapable to those committed to this community. Outsiders however will regard it as the outcome of a private and arbitrary choice that can not be justified to those who did not make this choice. The neutral common language however is both acceptable and imperative to every individual, regardless of his personal commitments towards particular moral perspectives.

In Engelhardt's view, pluralism has at least one disadvantage. The more the moral communities within a pluralistic society differ, the more impoverished and regulative the neutral common language will have to be. Indeed, in pluralistic society, ethics tends to give way to regulation. Where communities do not share a common moral sense, detailed regulations and a bureaucratic idiom are inevitable. Compared to the moral idiom of a particular moral community, the neutral common language seems a very impoverished substitute. Therefore, in particular moral communities the individual still has to seek those moral answers the neutral common language can not provide. Pluralism implies the decline of moral content in the public realm. This disadvantage is the prize we have to pay for living in a pluralistic society.

By endorsing a neutral common language for moral deliberation we indicate that we prefer peaceful negotiation, tolerance and resolution by agreement to moral warfare. Those who participate in moral deliberation must respect the autonomy of other participants. Those who do not respect the autonomy of others, disqualify themselves as participants and excommunicate themselves from the scene of moral deliberation. Indeed, the concept of autonomy is a crucial item in the vocabulary of the neutral common language.

In bioethics, the minimalist view on moral deliberation in pluralistic society is widely endorsed, but also disputed. A prominent opponent of this view is Daniel Callahan. According to Callahan, minimalist ethics tends to confuse useful principles for determining the relationship between the individual and the state with the broader requirements of moral life. It provides us with a thin and shriveled conception of morality and deprives us of a meaningful language to talk about our life together outside of our contractual relationships [3]. Moreover, the so-called neutral idiom of minimalist ethics is not neutral at all, but contains a particular morality of its own (i.e. liberalism). Being one particular moral idiom among others, it does not overrule its rivals. According to Callahan, it is not the purpose of ethics to maintain social peace, but to tell right from wrong in difficult health care situations. Sometimes social peace must be thoroughly disturbed on behalf of moral values. Finally, the language provided by contemporary ethics seems to be the artificial language of a particular moral community (the community of ethicists) that does not prove very adequate or useful outside this rather sealed world of professional ethics; it falsifies the richness of moral life by reducing real life dilemma's to technical problems.

To elaborate Callahan's objections, the following argument could be made. The neutral common language, provided by bioethics, is not neutral at all but, rather, the moral language of a particular community: the community of bioethicists. If the neutral common language is the language of a particular community, this language will *seem* binding and inescapable to those that participate in this community (i.e. bioethicists). To others however, to those who are not involved in this particular community, the "neutral common language" might seem as parochial and arbitrary as other moral idioms in pluralistic society. From the point of view of other moral communities, bioethics does not coin a neutral moral language, but merely adds one *particular* moral language to those already available. To those involved, the neutral common language, the idiom of autonomy and consent, seems inescapable, to outsiders however it will not. Why, then, should the so-called neutral common language, provided by bioethics, overrule the vocabularies of other communities? Why should the vocabulary of bioethics be considered binding where other moral vocabularies are considered to be ultimately arbitrary? Why should this particular moral language be accepted by the native speakers of other moral languages as their "first language"? In short, minimalist ethics raises the suspicion of being a "coup d'état" of one particular moral idiom at the expense of others. Furthermore it seems that, in Engelhardt's proposal, the problem of pluralism, of incommensurable views and irresolvable differences, is rather avoided than resolved. Serving as a congenial – according to Callahan a rather too congenial – vehicle for moral dispute, the neutral common language itself is considered indisputable.

Thus, in bioethics, minimalist ethics is widely endorsed to overcome the threat of moral warfare posed by pluralism. However, several drawbacks and disadvantages of the minimalist solution have been identified as well. To emphasize the significance of this debate, I will now turn to a controversy in moral philosophy at large that, up to a certain point, will be recognized as similar. I will briefly refer to the work of Apel and Habermas in Germany who share Engelhardt's concern with consensus as the outcome of peaceful deliberation and who elaborated a "discourse ethic" that in some important respects can be considered akin to minimalism. Next, I will introduce the French philosopher Lyotard as an antagonist of the German discourse ethic. Finally, I will refer to MacIntyre and Stout to focus attention once again on the quest for a neutral moral language as a vehicle for moral deliberation, as advocated by Engelhardt.

IV. CONSENSUS: ENDORSED BY APEL AND HABERMAS, CHALLENGED BY LYOTARD

One important reason for introducing Apel and Habermas is, that these philosophers deny that pluralism is an inevitable condition of contemporary moral deliberation. I will first turn to Apel and then to Habermas.

Unlike Engelhardt, Apel rejects the watershed between – on the one hand – the neutral common language of pluralistic society and – on the other – the particular moral idioms or perspectives of particular moral communities *within* this pluralistic society [1]. According to Apel, particular moral communities no longer furnish adequate answers to the global questions that present themselves to us. Every individual who contributes to the moral debate addresses himself to, and becomes a participant in, the *universal community* of participants in moral deliberation.

In spite of this difference, Apel clearly shares Engelhardt's concern for the ethics of deliberation. By committing ourselves to the community of participants in moral deliberation, we acknowledge the ethical principles implied: mutual recognition and respect among participants. By arguing we express a "will to argue", a willingness to endorse the ethics of argumentation. The speech-acts of those who criticize the ethics of argumentation are inconsistent, because they criticize the very principles they tacitly acknowledge by taking the floor. The ethics of argumentation are imperative to every participant in moral deliberation. They are independent of commitments of participants towards particular moral communities or perspectives.

However, the following objection could be made. If we agree that every individual who argues about morality addresses himself to a community of participants in deliberation, this does not necessarily imply that he addresses himself to a *universal* community, encompassing every human being who argues about morality, as Apel would have it. He might merely address himself to one of the particular moral communities Engelhardt has in mind, ignoring those participants who do not share his basic moral assumptions. We might clarify this objection by asking what language will be spoken in Apel's community of participants in moral deliberation. In contemporary society, several moral languages are available. Therefore, a hermeneutic attitude seems imperative. Before advocates of different moral communities can argue, they have to "understand" each other. A hermeneutic attitude however, is clearly rejected by Apel. He acknowledges that hermeneutics might contribute to the moral sensitivity of individuals, but emphasizes that this might result in individuals becoming too sensitive towards other moral languages and this in turn might result in relativism and a lack of moral resistance towards threatening perspectives and idioms that reject peaceful negotiation and deliberation and opt for warfare, like nazism and anti-semitism.

Apel's perspective is quite congenial with the views of Jürgen Habermas. Habermas shares Engelhardt's and Apel's concern to elaborate an ethic of moral deliberation, oriented towards consensus formation, in which no coercion, force or pressure is at work apart from the persuasive force of reason and argument, and in which every participant is willing to disclose reasons for his moral decisions and convictions, that is: to refrain from any appeal to subjective intuition or dogmatic conviction ([5,6]). Habermas realizes that this ideal of moral deliberation will never be fully realized in real-life discursive situations. Yet, although it is undoubtedly a philosopher's fiction, we can consider it a "regulative idea" that will have an impact on reality insofar as we try to live up to it and use it as a critical tool for judging particular instances of deliberation.

In a recent contribution Habermas [7] challenges the claim that, in contemporary society, several incommensurable but equivalent moralities coexist; he thus challenges the very idea of pluralism. Furthermore, he challenges the idea that the particular moral perspectives of particular moral communities should compensate for the loss of moral content in the public realm. To support this view, Habermas, very much like Apel, refers to the ethics of moral deliberation. While arguing, Habermas claims, we aim at mutual understanding among participants. We are not doomed to settle for the particular moralities of particular moral communities. We might criticize these moralities without being unintelligible to those committed to them. In a situation of perfect communication (which is bound to remain an ideal but which nevertheless can be approximated in actual communication) there are no considerations, views or arguments which cannot be made an item of debate. Arguments which are intelligible or even convincing here and now, to this particular audience, can be explained to others as well. We might summarize Habermas' discourse ethics by saying that the concept of pluralism is rejected, the concept of consensus formation endorsed. The agenda of deliberation might contain any conceivable item, yet every item must be made intelligible and put to the test of unrestrained deliberation.

Lyotard disagrees with Habermas' discourse ethics both on the issue of pluralism and on the issue of consensus formation. He endorses the concept of pluralism, but rejects the concept of consensus ([8,9]). He claims that Habermas' conceptual denial of pluralism and his elaboration of the concept of consensus, violate the actual pluralism that characterizes postmodern society. Lyotard aims at elaborating a view on social discourse that will sustain true pluralism and yet counter the threat of moral warfare.

According to Lyotard, we must distinguish several language games within social discourse, such as: the prescriptive language game of ethics (directed at justice), the denotative language game of science (directed at truth), the performative language game of regulation and control (directed at efficiency) and the narrative language game of meaning and identity (directed at religious or metaphysical pictures of human life). At present, the performative language game tends to subordinate and dominate the others. The prescriptive language game of ethics for instance, tends to be subordinated to the principle of performativity, that is: it tends to be considered of value only insofar as it contributes to the development of efficient procedures for dealing with moral problems in society. This development must be countered by stressing the incommensurability of the prescriptive and the performative language game. Both language games should be respected, and allowed to pursue their proper aims. A situation where one particular language game subordinates another is referred to by Lyotard as "discursive terror".

Although Habermas' discourse ethic is meant to be a counterforce to these terrorist tendencies of the performative language game, the solution offered by Habermas is inadequate. According to Lyotard, Habermas' consensus ideal assumes that participants in deliberation will agree upon the rules of the language game played in deliberation. This assumption is at odds with the postmodern principle of the plurality of language games and, by implication, with the plurality of discursive rules encountered in postmodern social discourse. Moreover, Habermas' discourse ethic is far from neutral. In his writings, Lyotard time and again tries to show that Habermas' ethic is embedded in a narrative language game. ultimately, he appeals to one of the grand narratives of Enlightenment: the emancipation-narrative, drawing a picture of the continuous progress of mankind towards articulacy and autonomy. In a postmodern context however, these grand narratives have lost their credibility. They are no longer acceptable or intelligible to all participants in deliberation. Furthermore, Habermas' assumption that participants may reach agreement on the rules for deliberation is a rather tricky assumption, as it might expose social discourse to the terrorist tendencies of the language game of efficiency and regulation, for this is the language game that time and again urges itself upon participants in acute discursive situations. Lyotard's plea for postmodernism, as opposed to Habermas' adherence to modernism, is a plea for true pluralism. Postmodern pluralism is threatened both by procedural tendencies in ethics (that try to subordinate the quest for justice to the demands of efficiency and regulation) and by discourse ethic (that denies the incommensurability of language games in contemporary society).

Let this suffice for an account of the debate between discourse ethic and postmodernism in contemporary philosophy. If we briefly compare Lyotard's postmodernist view with the minimalist view of Engelhardt it is clear that both advocate the pluralism of postmodern society. While Lyotard emphasizes the incommensurability of the prescriptive language game (directed at justice) and the performative language game (directed at efficiency and regulation), Engelhardt emphasizes the incommensurability between the prescriptive language game of ethics and the narrative language game of broader moral and religious commitments. At this point however, there is also an important difference between Lyotard and Engelhardt. Lyotard would consider Engelhardt's account of ethics much too similar to the performative language game of regulation. Although Engelhardt claims he respects pluralism, from a Lyotard-like perspective it is rather worrisome that in a pluralistic society ethical discourse tends to give way to regulation. Lyotard would consider this a submission of the prescriptive language game (seeking justice), to the performative language game (seeking efficient regulation). For this reason, Lyotard would consider the decline of ethics into regulation a threat to pluralism, a "coup d'état" of one particular language game at the expense of others. Lyotard would claim that Engelhardt's neutral common language tends to obey to the performative imperative (aiming at efficient procedures to deal with moral conflicts) rather than to the prescriptive imperative (aiming at justice).

From what has been said above we may conclude that the problem of pluralism, central to Engelhardt and Lyotard, seems somewhat underestimated by Apel's and Habermas' optimism. Engelhardt emphasizes that bioethical questions arise "against the backdrop of a moral crisis". And although Callahan considers the claim that our societies are pluralistic beyond repair unproven, he himself points to the dispute over abortion as a chronic moral difference: participants do not seem able to convince each other. It is a case of conflict between incommensurable moral perspectives. The same is emphasized by Alasdair MacIntyre who, in *After virtue*, mentions abortion as a paradigm example of the interminable and insolvable character of contemporary moral disputes. MacIntyre, a prominent voice in contemporary moral philosophy, deplores and emphasizes the harms of pluralism. He would not be satisfied with Apel's or Habermas' proposal to put all moral arguments and convictions to the test of free and unrestrained deliberation. In the following section we will focus attention on his views and then turn to a critical review of them by Jeffrey Stout.

V. DEPLORING AND ENDORSING PLURALISM: MACINTYRE AND STOUT

In *After virtue*, MacIntyre points to the state of utter confusion in which contemporary morality finds itself. Moral disagreement seems to be insolvable and interminable by argument. There seems to be no way of securing rational agreement in our culture [10]. Pluralism, or the failure to provide a single uncontestable moral viewpoint, is considered an undesirable condition, quite unfavourable for moral deliberation. Indeed, MacIntyre considers deliberation among moral strangers pointless. Instead, he recommends individuals to retreat into small, counter-cultural communities of those who consider themselves morally akin, in expectation of a moral era less obscured and confused than ours.

From the point of view of minimalist ethics the following question arises: does this imply that MacIntyre, in case of moral conflict among moral strangers, will opt for moral warfare instead of moral deliberation? It is my contention that he does not. MacIntyre's proposal does not stand a chance unless the peaceful coexistence of moral communities is warranted. Although as a rule we might refrain from substantial moral dialogue with strangers, in case of a moral conflict we will find ourselves forced to participate in peaceful negotiations. A resolute refusal of deliberation will inevitably bring about a condition of moral warfare in which MacIntyre's moral communities will not be able to maintain themselves. Spokesmen of particular communities therefore must comply to the moral demands of minimalism, i.e. willingness to participate in moral deliberation, mutual respect among moral strangers and the endorsement of a neutral moral language to settle conflicts with the world outside. MacIntyre's portrayal of contemporary society has been reviewed extensively by Jeffrey Stout. In the following, I will agree with Stout's claim that MacIntyre's narrative is penetrating, but ultimately unconvincing.

In accordance with MacIntyre, Stout emphasizes that contemporary culture is marked by disagreement [13]. Unlike Macintyre however, he reconciles himself with pluralism instead of deploring it. The title of his book is significant: *Ethics after Babel. The Languages of Morals and their Discontents*. It indicates the prominence of the language metaphor of pluralism. According to Stout, we encounter many moral differences in public deliberation. Some go beyond mere disagreement over the truth of particular propositions. They point to the existence of *distinct moral languages* in contemporary society. Indeed, in contemporary moral discourse, several moral languages are involved. Too many, according to MacIntyre who, as Stout puts it, assigned himself the role of a Biblical prophet deploring the present state of things.

Unlike MacIntyre, Stout does not *deplore* pluralism. On the contrary, he criticizes philosophy's urge to develop a universal language in which all moral differences can be stated and resolved. He defines this urge as the quest for a *moral Esperanto*. Engelhardt's "neutral common language" might count as a contemporary and liberal version of this quest. Against the longing for a uniform and unconfused moral idiom, Stout emphasizes the significance of disagreement and diversity, not mere disagreement over moral propositions, but fundamental disagreement between distinct moral languages in pluralistic society, between heterogeneous and incompatible schemes of moral justification.

This does not imply, however, that Stout calls for moral warfare. He acknowledges that secular moral discourse – from which minimalist ethics is derived – has emerged as an attempt to minimize the dramatic consequences of religious warfare and conflict that pervaded the dawn of modern times. Stout's option is a rather modest one. He endorses and advocates the language of minimalist ethics, of rights and tolerance for pragmatic reasons, as the best option available under the circumstances. In view of the threat of moral warfare, *phronesis* will make us side with a minimalist solution. Although it is far from being a *neutral* moral language, to accept it seems the wisest thing to do, not as a moral Esperanto, but as an *overlapping consensus*, as the outcome of creative moral "bricolage" which knows how to make incompatible vocabularies work in concert. Notwithstanding his discontent in contemporary moral discourse,

Stout does not side with those who attempt to restore religious moral idioms. Stout rejects both communitarianism and Esperantism, in order to endorse the language of rights, autonomy and consent as the most sensible option, because it proves to be the option that most people are willing to accept. MacIntyre portrays pluralistic society as too fragmented to sustain rational moral discourse. According to Stout, MacIntyre underestimates the consensus actually provided by the language of rights, tolerance and consent.

VI. REDUCED PLURALISM AND MORAL SUBJECTIVITY

In the preceding sections we reviewed several attitudes towards pluralism and consensus formation in contemporary ethics. From the point of view of minimalist ethics it is claimed that, in order to avert moral warfare, all participants in moral deliberation must endorse the three moral demands of minimalism. Yet, by some this encroachment on moral debate is being challenged. Some authors claim that attempts to introduce a neutral moral language to settle moral disputes, will conceal moral differences rather than resolve them. Especially the third moral demand of minimalist ethics, urging us to refrain from any appeal to particular moral intuitions or conceptions of life, and calling upon us to avoid the use of particular moral vocabularies in public deliberations with moral strangers who can not be expected to share these vocabularies, does not seem acceptable to all. This final section is an effort to formulate and answer three basic issues that emerged during the discussion in the preceding sections: Is contemporary society really pluralistic? Is pluralism to be advocated or to be deplored? Does minimalism sustain or rather impede pluralism?

One way of answering the question whether contemporary society should really count as pluralistic, is by means of historical comparison, for instance by comparing our moral condition to the medieval one. Everyone only slightly acquainted with the vast literature on medieval life will admit that *de facto* medieval society was tremendously pluralistic, heterogeneous, variegated and versatile and that, if anything, it was a society of conflict and divergence. Yet, in a very basic manner it might be claimed that medieval life was much more homogeneous than ours. All individuals shared the same religion, the same basic frame of reference, and apparently they shared the same basic experiences of life. On the intellectual level, for example, all individuals spoke the same language –

vulgar Latin – were bewildered by the same intellectual problems and participated in the same debates. This basic moral coherence of medieval life was summarized in the summit of medieval thought – the *Summa Theologiae* of Thomas Aquinas [2]. According to Thomas, a limited set of moral goals can be identified which are pursued by all individuals, while the final moral goal of man – happiness – is basically similar for everyone. And although several choices can be made and several options are available, all these choices and options adhere to the same basic pattern of moral life. In others word, the *de facto* diversity of medieval life was grounded in one basic form of moral subjectivity, shared by all individuals.

The sixteenth century, however, witnessed the rise of strong centrifugal tendencies in moral life. It witnessed the rise of unprecedented and incompatible forms of moral subjectivity, giving rise to the nation state on the political level, the formation of national languages on the cultural level, and the launching of the Reformation and the Counter-Reformation on the ideological level. In short, this emergence of incompatible forms of moral subjectivity resulted in true pluralism, which, as a moral condition however, proved to be short-lived. Before long, these incompatible moral forces clashed and the euphoria of Renaissance life was silenced by the violence of religious warfare and the subsequent efforts to counter the threat of warfare by increasing reliance on state power on the political level and, on the intellectual level, on the universal and homogeneous truth of reason, to be accepted by all, regardless of the ideological affinities or the vernacular spoken by the individual involved. A new form of moral subjectivity emerged, exemplifying the idea of modern citizenship and encouraging the development of modern sciences (like, for example, medicine).

At present, however, we witness what can be considered as the very reversal of the developments just described, at least in Europe: we witness the sudden decline of the nation state on the political level, of the national languages as a medium for scientific exchange on the cultural level, and of the significance of political or religious denominations on the ideological level. And finally, on the ethical level,we are faced with a decisive Either/Or. *Either* we opt for *true* pluralism, something similar to the pluralism which existed during the sixteenth century; *or* we shrink back from this possibility, since it entails the risk of moral warfare, and seek refuge in new forms of centripetal morality, symbolized by the introduction of a new *lingua franca* for public dispute. In the latter case, we would opt for what could be referred to as *reduced pluralism*. It would imply that, whereas true pluralism ('moral warfare') is to be deplored, reduced pluralism is to be encouraged.

But what would reduced pluralism amount to? And why should one be at all reluctant to endorse it? At first glance, it merely seems to imply that, although we are forced to accept a reduced vocabulary and a reduced agenda for public debate, every individual will be allowed to pursue his or her personal moral goals in private (or in the context of a particular moral community, if he or she is so fortunate as to find one). Eventually, however, we come to recognize that reduced pluralism entails a profound infringement on moral life. For it means that we have to become a different kind of person, someone willing and able to accept the basic *compartimentalization* of moral life into a public and a private sphere. That is, we will have to adhere to a certain kind of moral subjectivity. Reduced pluralism – or minimalism – implies that, notwithstanding the empirical variety of contemporary society in terms of life style, all forms of moral existence have to be grounded in one and the same basic form of moral subjectivity – the compartimentalized Self. This is the basic moral structure we have to accept, the kind of moral subject we have to become. And this implies that reduced pluralism, or minimalism, is far from being neutral. Whereas some forms of moral existence are encouraged, others are rejected or at least impeded, since they are considered as being incompatible with the basic form of moral subjectivity presupposed by minimalism which I (here and elsewhere) referred to as the compartimentalized Self [14].

The actual variety of postmodern moral life presupposes centripetal pacification on a basic level, and as such its differs form what I referred to as 'true' pluralism. Unlike what is suggested by Engelhardt, reduced pluralism does not grant individuals the right to proceed with their idiosyncratic ways of life, for they have to transform themselves in order to meet the demands of minimalism – they have to be transformed into a compartimentalized Self. And this is much closer to a conversion than to merely accepting an external set of temporary and provisional procedures. The speech acts of those who refuse to be transformed, are denied access to moral deliberation. When Engelhardt claims that all moral communities are to be considered as experiments in the building of moral worlds, we must keep in mind that all such experiments have to comply with, and fit into, a basic moral format. Some communities or forms of moral life will be more adapted to the basic moral demands of minimalism than others, and are therefore bound to be more successful in a reduced pluralistic environment. Callahan is right in emphasizing that minimalism is far from being neutral, for it rather entails a basic and substantial view of moral life and human relationships. Subsequently, Apel and Habermas clearly acknowledge that reduced pluralism, although it might allow for a broad range of individual variety on an empirical level, presupposes on a more basic level a universal form of moral subjectivity to be imposed on all participants in deliberation. And for this reason, Lyotard rejects their view as entailing moral violence, or even "discursive terror". In short, minimalism is not merely a matter of procedures. Rather it entails the imposition of a basic universal form of moral subjectivity – thus putting an end to "true" pluralism.

This article was about recognizing the true, and really tremendous impact of the Either/Or facing us in the present situation, rather than presenting a final solution or adhering to an ultimate decision. Stout already recognized the impact of the Either/Or described just now, but for reasons of prudence he finally reaches a conclusion which can be considered as rather similar to the one already drawn by Engelhardt. For although what Engelhardt referred to as a *neutral* common language is designated as a moral Esperanto by Stout (thus avoiding the claim that any moral language could be 'neutral'), they both eventually settle for what I referred to as *reduced* pluralism, a position which relies on the moral potential of the language which, due to the historical developments described above, happens to dominate contemporary moral discourse. Before making a final choice regarding this basic Either/Or (something which I actually tried to do elsewhere [14]), the moral impact of both options – of *true* versus *reduced* pluralism – are to be thoroughly assessed.

BIBLIOGRAPHY

1. Apel, K-O.: 1973, 'Das Apriori der Kommunikationsgemeinschaft und die Grundlagen der Ethik. Zum Problem einer rationalen Begründung der Ethik im Zeitalter der Wissenschaft', in K-O Apel, *Transformationen der Philosophie*, Vol. 2: *Das Apriori der Kommunikationsgemeinschaft*, Suhrkamp, Frankfurt am Main.
2. Aquinas, Th.: 1922, *Summa Theologica*, P. Marietti, Taurini.
3. Callahan, D.: 1981, 'Minimalist ethics: on the pacification of morality', in A.L. Caplan and D. Callahan (eds.), *Ethics in Hard Times*, Plenum Press, New York/London.
4. Engelhardt, H.T.: 1996, *The Foundations of Bioethics*, (2nd. ed.), Oxford University Press, Oxford/New York.

5. Habermas, J.: 1981, *Theorie des kommunikativen Handelns*, Suhrkamp, Frankfurt am Main.
6. Habermas, J.: 1983, 'Diskursethik – Notizen zu einem Begründungsprogramm', in J. Habermas, *Moralbewußtsein und kommunikatives Handeln*, Suhrkamp, Frankfurt am Main.
7. Habermas, J.: 1986, 'Die Einheit der Vernunft in der Vielfalt ihrer Stimmen', in J. Habermas, *Nachmetaphysisches Denken*, Suhrkamp, Frankfurt am Main.
8. Lyotard, J-F.: 1979, *La condition postmoderne. Rapport sur le savoir*, Les Éditions de Minuit, Paris.
9. Lyotard, J-F.: 1984, *Le différend*, Les Éditions de Minuit, Paris.
10. MacIntyre, A.: 1984, *After virtue. A Study in moral theory* (2nd edition), University of Notre Dame Press, Notre Dame.
11. Stanley, J.M. *et al.*: 1988, *The Appleton Consensus: Suggested International Guidelines for Decisions to Forgo Medical Treatment* (Study Edition), Lawrence University, Appleton.
12. Stanley, J.M. *et al.*: 1989, 'The Appleton Consensus: Suggested International Guidelines for Decisions to Forego Medical Treatment', *Journal of Medical Ethics* **15**, 129-136.
13. Stout, J.: 1988, *Ethics after Babel. The Languages of Morals and their Discontents*, Beacon Press, Boston.
14. Zwart, H.: 1996, *Ethical consensus and the truth of laughter. The structure of moral transformations*, Kok Pharos, Kampen.

AKIO SAKAI

CONSENSUS FORMATION IN BIOETHICAL DECISIONMAKING IN JAPAN: PRESENT CONTEXTS AND FUTURE PERSPECTIVES[1]

I. INTRODUCTION

Today, in Japan, arguments concerning the introduction of new biomedical technologies, especially those related to birth and death, the "edges of life", focus on the process of consensus formation. For instance, heart transplants (first attempted in 1968) have been discontinued for about 30 years, not because there was no demand for them, or because medical or technical problems were insurmountable, but because there was no positive *consensus*, especially public consensus, concerning the claim that total brain death is equivalent to the death of the individual (in Japan, "brain death" requires death of the entire brain including the brain stem). Though "social agreement" or "public consensus" is a vague and not well-defined notion, it has become, as S. Segawa put it, "the key-word when talking about the problems of brain death and organ transplants in our country" ([27], p. 93).

Aside from public consensus, occasions have been increasing in which physicians and scholars from various specialties have gathered to discuss various ethical problems in biomedicine. Such efforts are no doubt contributing to a mutual understanding among scholars, and the quality of argument for achieving consensus by discussion has improved. However, it is also true that there is a critical attitude regarding the significance of these meetings, as expressed, e.g., in the remark that "medical doctors, biologists, philosophers, and theologians [are] to meet together and to listen to such clichés as 'it is necessary that people from various fields talk with one another and continue actively to obtain consensus' with a feeling of satisfaction, and then they go home" ([33], p. 226). The problem concerns the levels of consensus formation by discussion, and how to proceed from these various levels to an *academic consensus*.

In the following, I shall describe (1) the present situation in Japan and the problems of 'consensus formation by discussion', and (2) the notion of public consensus (which is very influential in Japan) in the context of

H.A.M.J. ten Have and H.-M. Sass (eds.), Consensus Formation in Healthcare Ethics, 93–106

the introduction of novel biomedical technologies, especially the diagnosis of brain death and the procedures involved in organ transplantation.

II. LEVELS OF CONSENSUS FORMATION REGARDING BIOMEDICAL ISSUES

Consensus formation occurs in different contexts, e.g., *ad hoc* committees, academic societies, hospitals and other institutions. A number of *ad hoc* committees, each of which was organized for a particular purpose, have made public the results for which they obtained a consensus. For example, the Committee for Brain Death of the Japan Association for EEG made public their criteria for the diagnosis of brain death. In December 1985, a research team on brain death of the Ministry of Public Welfare published new criteria for the diagnosis of brain death. In May 1986, the committee on brain death of the Japanese Association for Legal Medicine published an interim report stating that "[total] brain death can be considered the death of the individual." On January 12, 1988, the Bioethics Council of the Japan Medical Association issued a final report on brain death and organ transplantation which clearly encourages organ donation [17]. However, in some cases, controversy was too intense to establish a consensus even within, for example, the Statesmen's League for Research on Bioethics, which withdrew its interim report supporting brain death criteria in the determination of death of an individual.

Within academic societies, consolidated views have been obtained and announced, e.g., the Japan Attorney's Association criticized the Bioethics Council of the Japan Medical Association that supported the use of brain death criteria, and in December 1986, the Japan Association for Transplantation set forth its policy on organ transplants from brain death patients [11]. In 1983, the Japan Association for Gynecology and Obstetrics published its opinion concerning IVF-ET.

The most suitable medium for establishing a substantial "consensus" among physicians and scholars from various specialties in institutions is the ethics committee (EC). These have been formed in each of Japan's medical schools. The ECs at these medical schools have been established to arrive at an *ethical consensus* at the level of the institution. This is especially the case concerning new methods of treatment and the initiation of clinical trials. Members of the ECs not only include professors in the medical schools, but professors of law and literature as well.

However, there is still criticism of the structure and function of these rapidly established ECs. T. Saito has indicated that many of the ECs at medical schools are not actively working and have not yet moved along the right lines; moreover, the establishment of hospital ethics committees has much to contend with [24]. As to the significance of ECs themselves, it has been argued that they do not clarify where various responsibilities lie. In addition to that, N. Sawada has expressed concern that, "though an appeal to an EC can be admitted as the second best policy, doesn't it tend to lead to physicians' ceasing to be moral themselves, becoming bureaucratic, and abandoning their pride which they once had as physicians?" [26].

S. Yonemoto has sharply criticized the attitude of medical professionals, contending that they have only borrowed the results of discussions in the West in which the Judeo-Christian view of family and body is strongly reflected, and that they have evaded public objection by attaching some restricting conditions reflecting the Japanese situation, to that result ([33], p. 219).

In spite of the various criticisms concerning the structure, quality, proper activity, functions, and significance of ECs, the necessity of having ECs is generally accepted. Moreover, in December 1988, *The Liaison Society for Ethics Committees at Japan's Medical Schools* was founded as a national, deliberative body that could relate all ECs to one another and assist communication between them [1]. Though the activity of ECs differ at each medical school, there has been one case at the EC at Tokushima University where the introduction of IVF-ET was discussed with the *full participation of humanists representing various specialties.* The ECs at the medical schools and the *Liaison Society* continue to remain the most appropriate mechanism for establishing an *academic consensus* concerning the application of new biomedical technologies.

III. SOCIAL AGREEMENT AS PUBLIC CONSENSUS

The terms 'social agreement' and 'public consensus' have frequently been referred to in discussion of the ethical aspects of biomedicine, and "recently has been worn out like a dustcloth" [4]. For example, the typical argument proffered by those who are cautious concerning the introduction of total brain death criteria in the context of organ transplantation is that: (1) the clinical application of the criteria for brain death requires public consensus; (2) public consensus has not yet been established in this matter; (3) thus total brain death criteria cannot be clinically applied.

However, the problem with this formulation is that the phrase "public consensus has not yet been established" is not clearly understood. Moreover, it is also unclear whether or not public consensus can be achieved, or what should count as the establishment of public consensus. To understand the function of public consensus, one should look back over the course of argument up to the final approval of the Organ Transplant Bill (including total brain death criteria) of June, 1997. In the late 80s, those who wished to promote the use of total brain death criteria were becoming more and more impatient. In the final report concerning brain death and organ transplantation of the Bioethics Council of the Japan Medical Association, one finds the following remark:

> There is a group of people who strongly argue that social agreement or consensus of the nation is necessary to allow for the determination of death by [total] brain [criteria]. From this point of view, it is still premature to allow the determination of death by [total] brain death [criteria] without social agreement. How can this social agreement be established and confirmed? Generally speaking, the argument for public consensus on the part of those who insist that a determination of death by [total] brain death [criteria] is premature, is nothing but the expression of the feeling that the argument by the great majority of the nation is required, but has failed to show the factors and procedure that is required to ascertain what social agreement is and how it is to be confirmed. Such an "argument" for public consensus only leads to bypassing the problem and letting it remain unclear [5].

Often, the results of public opinion surveys were taken by both sides (the positive and cautious standpoints) as proof of whether or not a public consensus had been established. However, this is surely inadequate. For in these surveys each item on the questionnaires allows for some interpretation. Moreover, not only the quality of such questionnaire but the basic knowledge of respondents have been criticized. T. Tachibana notes that biased results contaminated on survey because the questions were too leading. Besides that, he casts doubt on the respondent's knowledge of the brain death criteria and the significance of such surveys ([29], pp. 56-57).

Since there is no standard available to determine the correlation between a tally and the establishment of a consensus, the results of such surveys can be interpreted in quite different ways, depending on the data that is stressed. In short, since there is no logical relationship between the results of such surveys and public consensus, the conclusion based on the results is inevitably ambiguous. What surveys tend to reveal is not consensus, but rather the *degree of differences of opinion* reflected in specific items and public opinion trends.

In order to end the search for public consensus, some demanded a legal decision concerning the problems introduced by brain death criteria [31]. However, there was general agreement that public consensus was necessary for such legislation. Since public consensus is considered the important standard by which to approve a particular social behavior *prior* to any legislation, the conclusion was that it was impossible to legislate without public consensus. H. Mizutani remarked in 1986:

> Perhaps it is still impossible in Japan to realize legislation that is the concrete method of making society accept [total] brain death criteria, though it has already been realized in the U.S. . . . Public opinion has a dominant influence in Japan. When public opinion is negative, legislation is rarely proposed ([20], pp. 296-297).

Consequently, it took over ten years until the legislation was approved; indeed the debate continues. It is obvious that "the requirement of public consensus" functions as an objection to the introduction of the notion of total brain death and organ transplantation. Is this due to the fact that the Japanese pursue a *consensus gentium* (unanimous consent as the standard for truth)? Or is there any particular factor at work in the "public" that resists the application of new technologies? The answer is revealed in the discussion concerning the following question: "Is it possible or impossible to establish a public consensus?"

The typical argument of those who admit the possibility of achieving public consensus may be formulated as follows: First of all, on the part of medical professionals, (1) total brain death criteria must be strict and reliable, (2) physicians must be trained to unmistakenly diagnose total brain death, and (3) consensus should be established among medical professionals. In addition, medical professionals should enlighten lay people concerning brain death criteria. It will be necessary to abolish superstition and vulgar faith. When these conditions become satisfied, the possibility for establishing public consensus concerning total brain death and organ transplantation will become a reality.

This view appeals to rationality and the scientific attitude to assist in establishing public consensus. In other words, if we limit the discussion of the notion of total brain death criteria and organ transplantation to its rational, scientific aspects, public consensus will be established more naturally and more quickly. However, it is not only the small scientific circle, but *society in general* that must reach a consensus. Though it is important to provide basic scientific knowledge, the view that this alone leads immediately to a public consensus is mere "scientism" – the claim that scientific outcomes are *a priori* true and therefore should be accepted by everyone.

Public consensus is hard to formulate logically. It is, however, always reflected in our daily life. M. Polanyi remarks:

> I cannot speak of a scientific fact, of a word, of a poem or a boxing champion; of last week's murder of the Queen of England; of money or music or the fashion in hats, of what is just or unjust, trivial, amusing, boring or scandalous, without implying a reference to a consensus by which these matters are acknowledged – or denied to be – what I declare them to be ([22], p. 209).

Consensus formation functions to define our everyday life, and forms a framework of collective thought. "A new device has obtained public consensus" means that a new device has been assimilated into a particular society in which various cultural and unscientific components exist. Then the thesis – the appeal to the rational and scientific components and the elimination of conflicting factors is the only way to establish public consensus – reveals the force of non-scientific and normative factors, i.e., religion, emotion, and custom, as objections to the establishment of public consensus when introducing novel biomedical devices and other technologies.

The "requirement of public consensus" signals, at least partially the necessity for considering various cultural factors in the application of new technologies, and "public consensus" functions as the *symbol* of the consolidated will of the Japanese in which both scientific *and* cultural factors must be harmonized.

The actual establishment of consensus should be considered a turning point where discussions of consensus must be understood as reflecting a very *dynamic* process. Such a dynamic relation is reflected in any challenge to traditional culture, e.g., the Japanese view of life and death. A radical change in traditional thought which infiltrates the Japanese people is necessary if Japanese society is to accept novel biomedical technologies.

IV. CONSENSUS AND THE PRINCIPAL CHARACTERISTICS OF JAPANESE CULTURE

The Japanese are said to be an homogeneous people. It is often pointed out that there is no tradition for establishing consensus by debate and discussion, since Japanese society is a unitary society and the people *share* too many basic values. For example, H. Sakamoto remarks that while "in the western world the theme 'what constitutes consensus' has been refined in a stern history", in Japan 'what is consensus and how it is established' is not fully understood; therefore the notion of consensus itself has no roots in Japan ([25], p. 64).

It has also been said that the Japanese people are not only aware of but tend to preserve their homogeneousness, and at every level – family, district, nation ([30], p. 60). The "feeling of never being content unless *all* are involved and united together" ([14], emphasis by me) is often observed in discussions or debates. These observations make it reasonable to suppose that there exists a strong desire on the part of the Japanese to preserve homogeneousness of thought and behavior.

In relation to the desire on the part of the Japanese for homogeneousness, one can note two characteristics in the way the Japanese establish *consensus by discussion*: (1) a tendency not only to avoid emphasizing differences, but to speak eclectic solutions in which conflicting views may coexist, i.e. a tendency to "stressing the points of agreement, rather than illuminating the differences" ([9], p. 153), (2) a tendency that once a consensus has been established, it is firmly adopted and admired as an important norm that cannot be easily changed, i.e., "[i]n Japan, at the point a 'consensus' is established, every disputed point dies like a fossil. Those who recall the past are criticized as if they were violating Japanese ethics" ([13], p. vi). Probably the fact that "in Japan, a law is seldom amended after enacted" ([16], p. 41) has something to do with the latter characteristic.

Such characteristics function as negative factors when forming a decision or a consensus in debates concerning the issues essentially related to human life, birth, and death in biomedicine, i.e., the notion of total brain death, organ transplantation, and IVF-ET. Concerning these issues differences of opinion are quite conspicuous and too serious to compromise. Moreover, the importance of these issues leads to an extremely cautious attitude on the part of the Japanese people regarding consensus formation, because such a consensus is *liable to dominate the future choices of every individual.*

It is also important to point out that new biomedical technologies often conflict with Japan's traditional culture, including religious beliefs, views of life and death, and the dead person's remains. Thus, it is necessary to investigate the dynamic relationship that exists between the religious subconsciousness of the Japanese and all novel technologies. This dynamic relation is the most important factor not only for achieving consensus by *discussion* but also for achieving *public* consensus.

It is not an easy task to clarify the meaning of religion in Japan, where "modernity and tradition, rational and irrational forces, and a delicate sentiment are entangled" ([20], p. 300).

Based on the results of a survey showing that "only 16% of 187 physicians and 16.4% (499) of the general public believe in life after death", Y. Hara indicates that religion has lost its persuasive power among today's people ([8], p. 266). Nevertheless, according to Y. Ikemi "though only 20% [of the Japanese] practice their religion, 80% believe that they are directed by an invisible force" ([10], p. 346). While "it is said that most Japanese do not strictly believe in a particular religion", in their behavior and "in the religious ceremonies, there is an undercurrent of a unique religious character" [23]. A. Hirakawa has analyzed the religious characteristics of the Japanese. First of all, "what forms the Japanese people's view of life and their religious mind" is Buddhism united with Shinto. The characteristics of the ancient, traditional religious mind of the Japanese is "the rejection of mind and matter [body] dualism". For example, Hirakawa has noted some religious factors that hinder the acceptance of the notion of total brain death and organ transplantation: (1) since the Japanese "do not like to distinguish between mind and body," and, tend to "bestow the same sacredness on body as on mind they tend to hesitate and do not accept [total] brain death as the death of the individual"; (2) the Japanese believe that "the spirit dwells in the remains"; and (3) "the Japanese funeral is a ritual for the repose of [human] souls." Given the Japanese "fear for the soul of the dead, especially the fear that the soul may become the evil spirit," they do not wish to injure the body ([9], p. 151-160).

The Japanese view of life and death, and the dead person's remains have frequently been referred to in the context of rejecting the introduction of the notion of total brain death in the context of the procedure known as heart transplantation ([34], p. 23). Since it is clear that the Japanese view of life and death is a critical component of Japan's culture, the pessimistic view remains that, as one physician put it, "it is impossible to alter the Japanese concept of death – which has a very long history – or to achieve a consensus among the entire nation" (see [32], p. 45). The Japanese view of the dead person's remains is also frequently mentioned , e.g., that – "the dead body is not simply inert matter, for it has consciousness and emotion, since the spirit dwells in it for some time" [6]. This belief is closely connected with "the unique sentiment the Japanese have for the person's remains" ([7], pp. 81-82). Many authors believe that such a sentiment or attitude that dislikes inflicting harm on the body ([18,23]), is an impediment not only with respect to *autopsy*, but with regard to advancing the idea of total brain death criteria in the context of *organ transplantation.*

Since this view of life, death, and the dead person's remains are psychologically shared by many Japanese, it has a power to dominate discussions. H. Sakamoto relates a very interesting episode: "Recently a Buddhist communion held a symposium concerning heart transplantation. Toward the end of the symposium, someone remarked that 'I may not be able to go to Buddha without my heart'. Since then, pubic opinion suddenly changed; at last, the public decided to object to heart transplantation" ([25], pp. 66-67).

What does this episode signal? Did the participants in the symposium abandon their independence and yield to the religious value represented by this person's remark? Of course it is impossible to draw a definite conclusion from this brief remark. We must also take into account the fact that this phenomenon occurred in a religious context.

Summarizing the discussion to this point, it is perhaps useful to refer to E. Namihira's account: She carefully explains why it is so difficult for the notions of total brain death and organ transplantation to gain acceptance among the Japanese.

> On the one hand, as secularization advances, religious concepts among people are becoming unclear, and they are unable to clarify their own conceptual understanding of human life and death. But on the other hand, it is very difficult to advance the argument, since traditional concepts remain in the form of emotion and sentiments concerning this crucial problem for society and the individual, i.e., life and death. The emotion and sentiment are difficult to express in a language that everyone can share. Moreover, the emotion and sentiment work together as a strong motive in determining human behavior. It seems that this situa-

tion makes discussions on [total] brain death and organ transplantation much more complicated ([21], p. 62).

V. CONCLUSION

It is an important and urgent task to establish a social consensus concerning new biotechnologies and their proper employment, especially the use of total brain death criteria in the context of organ transplantation. But as we have seen, there are various problems in seeking to achieve a consensus among various *ad hoc* committees and ECs through *discussion* alone. Even if the ECs at Japanese medical institutions should make decisions, they are never regarded as final, in so far as the achievement of a wider *public* consensus has such great importance in Japan. In the report of the Council for Life and Ethics, there are the following remarks: "Even though the method to establish a consensus by discussion follows a democratic procedure, there will be much difficulty in obtaining 'psychological' compliance in the context of modifying traditional images of life and death, which have been influencing people for a long time. Accordingly, the majority of the Council agreed that the more trustworthy way is education concerning the notions of life and death" ([19], p. 185).

Behind the opinion that public consensus has a crucial significance for the application of new biotechnologies, it is understood that such applications take place where the life-world (as the basis of all human living) encounters scientific understanding. Furthermore, the demand that various social restrictions be placed on scientific researchers cannot be simply dismissed as the response of an emotional and irrational opposition ([12], p. 82).

It may be useful to refer to S.F. Spicker's account in order to grasp our present situation. First, following the phenomenologists, he defines the 'life-world' as the social and historical world prior to any scientific or mathematical conceptualization, a world which "comprises the sum of mankind's involvement in everyday affairs". Then he suggests that "the scientific and theoretical world of contemporary biomedicine – which is the foundation for clinical medicine as well as the world of bioengineering and technological invention", has itself dramatically affected our life-world. Applying this approach to today's situation in Japan, we now experience a new moment in which a particular bioengineering and technological invention may itself, after some time, "be assimilated into the meaning of the life-world" ([28], pp. 206-208). We in Japan are faced with the decision whether or not we shall accept a gradual transformation of our life-world by forming a consensus concerning the introduction of total brain death criteria in addition to the traditional criteria for determining the death of the patient.

S. Yonemoto has indicated that "while Japanese society has unhesitatingly adopted almost all scientific technologies, it reveals a strong resistance to medical technology as it bears on birth and death, e.g., IVF and organ transplantation" ([33], p. 14). The reason is that the introduction of these new technologies inevitably leads to the radical alteration of our life-world. Though science offers a common language everyone can in principle share, it has become more and more alien to us in the very process of advancing itself. Even medicine, which is now understood as an experimental science, has detached itself from the life-world during its more recent history. F. Katayama has pointed out that medical practice and religion, which have been combined together, were separated during the modernization in the Meiji era (a period after the Edo era) ([12], pp. 145-146). As R. Kawakami remarks, "medicine has developed as a part of modern European science. Now there is a discontinuity between that and Japanese social consciousness" [15].

Certainly, the establishment of a public consensus in Japan may be realized where both the scientific dimension of medicine and the influence of Japanese traditional values, including religious consciousness, converge. While there remains a cautious attitude toward "the argument for public consensus", there also exist demands to further progress of biomedicine and biotechnology.

The consensus we must achieve, however, is not the "conventional" one that is based on either scientific rationalism *or* traditional values, but a *structural* one that "posits in the first place an insightful relation of the person to his situation" ([3], p. 492). In so far as it is true that "there is no ethics but the domination of rationalism in scientific and medical research" ([20], p. 135), we in Japan are compelled to reconsider traditional ethics and to discover or even create a new ethics that will be the basis for public consensus concerning the introduction of various novel scientific technologies. In order to introduce the products of biotechnology, that never cease to appear, we have to confront the task of harmonizing today's cultural relativism with universal and traditional human values. We live in an age when "the ethics of society must be created, though it is to some extent rooted in public agreement" ([20], p. 146).

The task of establishing public consensus must be accomplished through discussions among medical professionals, humanists, people from various disciplines, and the general public. The most suitable mechanism for initiating these discussions will in all likelihood be the ECs at Japan's eighty medical schools. Each EC will have to discuss total and perhaps neocortical brain death criteria in the context of organ transplantation; each will have to reflect the consensus of the local community. There is no other way to create a new ethics than to achieve a broad consensus through open public discussion.

Department of Neuropsychiatry
Iwate Medical University
Morioka, Japan

NOTE

[1] I wish to express my personal indebtedness to Stuart F. Spicker, Professor Emeritus, University of Connecticut Health Center, for his extremely helpful supervision and suggestions during the writing of this essay.

BIBLIOGRAPHY

1. Anonymous: 1988, 'A Preparatory Committee Was Held for Establishing The Liaison Society for Ethics Committees at Japan's Medical Schools', *Japan Medical Journal* **3375**, 99.
2. Aoyama, H.: 1987, 'Science/Technology and Society', in R. Nagamoto and S. Yonemoto (eds.), *Meta-Bioethics*, Nihon Hyoronsha, Tokyo, Japan, pp. 65-86.
3. Asch, S.E.: 1952, *Social Psychology*, Prentice Hall, Englewood Cliffs, N.J.
4. Bai, K.: 1985, 'Scientific and Ethical Aspect of Medical Practice and the Role of Law', *Nihon Ishikai Zasshi* **93** (6), 1027-1038.
5. The Bioethics Council of the Japan Medical Association: 1988, 'Final Report Concerning Brain Death and Organ Transplantation', *Nihon Ishikai Zasshi* **99** (2), 261-283.
6. Fujii, M.: 1989, 'The Characteristics of the Japanese View of Life and Death', *Igaku no Ayumi* **150** (5), 327-329.
7. Fukuma, S.: 1987, *Considering Brain Death*, Nihon Hyoronsha, Tokyo, Japan.
8. Hara, Y.: 1986, 'Life and Death – Through the Experience in Hospice', in T. Yoshitoshi (ed.), *Physician's View of Life*, Nihon Hyoronsha, Tokyo, Japan, pp. 261-270.
9. Hirakawa, A.: 1989, 'On the Japanese View of Life and Death' in T. Nakayama (ed.), *Brain Death and Organ Transplant: Why Transplants Can Not Be Done in Japan?*, The Simal Press, Tokyo, Japan, pp. 151-172.
10. Ikemi, Y.: 1986, 'Terminal Care and the View of Life and Death' in T. Yoshitoshi (ed.), *Physician's View of Life*, Nihon Hyoronsha, Tokyo, Japan, pp. 345-354.

11. Ito, M.: 1987, 'The Current Argument Concerning Brain Death', *Kango Tenbo*, **12** (10), 954-955.
12. Katayama, F.: 1989, 'If Conditions Are Fulfilled', in T. Nakayama (ed.), *Brain Death and Organ Transplant: Why Transplants Can Not Be Done in Japan?*, The Simal Press, Tokyo, Japan, pp. 145-150.
13. Kato, H. and Iida, N. (eds.): 1988, *The Foundation of Bioethics: Bioethics in Europe and the U.S.*, Tokai Daigaku Shuppankai, Tokyo, Japan.
14. Kato, I.: 1986, 'On New Brain Death Criteria', *Iryo* **2** (2), 26-27.
15. Kawakami, R.: 1989, 'Brain Death as a Legal Problem', *Igaku no Ayumi* **150** (5), 339-341.
16. Kawashima, T.: 1967, *Legal Awareness of the Japanese People*, Iwanami Shoten, Tokyo, Japan.
17. Kimura, R.: 1989, 'Anencephalic Organ Donation: A Japanese Case', *The Journal of Medicine and Philosophy* **14**, 97-102.
18. Manaka, S.: 1987, 'Determination of Brain Death', *Rinsho to Kenkyu* **64** (7), 2121-2126.
19. The Ministry of Public Welfare (ed.): 1985, *On Life and Ethics: The Report of the Bioethics Council*, Igakushoin, Tokyo, Japan.
20. Mizutani, H.: 1986, *The Argument of Brain Death: The Meaning of Living and Dying*, Soshisha, Tokyo, Japan.
21. Namihira, E.: 1988, *Brain Death, Organ Transplant, and Truth Telling about Cancer: The Anthropology of Death and Medical Practice*, Fukutake Shoten, Tokyo, Japan.
22. Polanyi, M.: 1958, *Personal Knowledge: Towards a Post-Critical Philosophy*, University of Chicago Press, Chicago, Illinois.
23. Sadamitsu, D. and Takeshita, H.: 1987, 'Social Problems Concerning Brain Death', *Kyu-Kyu-Igaku* **11** (7), 827-833.
24. Saito, T.: 1989, 'Ethical Considerations on Publication of Research Works – Responsibility of Editors and Institutional Ethical Committee, *Igaku no Ayumi* **150** (5), 321-323.
25. Sakamoto, H.: 1989, 'The Necessity to Reconsider Bioethics', in T. Nakayama (ed.), *Brain Death and Organ Transplant: Why Transplants Can Be Done in Japan?*, The Simal Press, Tokyo, Japan, pp. 61-68.
26. Sawada, N.: 1989, 'Evolution of Ethics', *Igaku no Ayumi* **150** (5), 307-9.
27. Segawa, S.: 1988, *The Actual Spot of Heart Transplant*, Shinchosha, Tokyo, Japan.
28. Spicker S.F.: 1982, 'The Life-World and the Patient's Expectations of New Knowledge' in W.B. Bonderson, H.T. Engelhardt Jr., S.F. Spicker and J.M. White Jr. (eds.), *New Knowledge in the Biomedical Sciences*, D. Reidel Publishing Company, Dordrecht, The Netherlands, pp. 205-215.
29. Tachibana, T.: 1986, *Brain Death*, Chuo-Koron-sha, Tokyo, Japan.
30. Tsuji, S.: 1989, 'Be Prudent to Accept Brain Death', in T. Nakayama (ed.), *Brain Death and Organ Transplant: Why Transplants Can Not Be Done in Japan?*, The Simal Press, Tokyo, Japan, pp. 55-61.
31. Uozumi, T. and Oki, S.: 1985, 'Concept of Brain Death', *Chiryogaku* **14** (4), 438-442.
32. Watanabe, T. (editorial supervision): 1988, *Studies of Death*, Niki Shuppan, Tokyo, Japan.
33. Yonemoto, S.: 1987, 'From the Viewpoint of Life Science', in R. Nagamoto and S. Yonemoto (eds.), *Meta-Bioethics*, Nihon Hyoronsha, Tokyo, Japan, pp. 217-241.
34. Yoshimura, A.: 1988, Presentation in the Panel Discussion in The Japanese Association for Artificial Organs (ed.), *The Replacement of Organs and the Revolution of Consciousness*, Asahi Shinbunsha, Tokyo, Japan, pp. 19-23.

ROBERT MULLAN COOK-DEEGAN

FINDING A VOICE FOR BIOETHICS IN PUBLIC POLICY: FEDERAL INITIATIVES IN THE UNITED STATES, 1974-1991

The recent prominence of bioethics in American government signals the increasing importance of medical and biological technologies in national life. It also shows the degree to which ethical analysis is recognized as essential in governing the use of new technologies. Some would add that the need for ethical committees in general, and in the federal government in particular, points out the erosion of shared values and the rise of moral pluralism, so that government commissions replace the traditional functions of a single church or other moral authority. In this light, federal commissions and grants programs can be seen as innovations to articulate common values and foster consensus in the void left by declining religious and cultural institutions.

The paper first reviews the history of the various commissions, advisory agencies, and programs in the United States. The purposes of federal sponsorship for bioethics are then discussed, followed by observations about the process of writing reports for the federal government. The paper concludes with a prescription for a balanced approach to bring ethical analysis into public policy, supporting both national commissions and university-based academic analysis.

I. HISTORICAL BACKGROUND

The National Commission for the Protection of Human Subjects of Biomedical and Behavioral Research (National Commission) was established with the signing of the National Research Act on July 12, 1974. The National Commission issued its first report, *Research on the Fetus*, in May 1975 [19]. In less than a year, the Commission met several times, commissioned lengthy background papers by outside experts, hired 16 staff, and drafted and revised the report many times. By late July 1975, the National Commission's recommendations had been translated into regulations. It was an impressive piece of work, presaging many more reports that laid a foundation for human subject protections in the United

H.A.M.J. ten Have and H.-M. Sass (eds.), Consensus Formation in Healthcare Ethics, 107–140

States. These in turn emerged as world standards.

This success story stands in stark contrast to the Biomedical Ethics Advisory Committee (BEAC), which crashed on the shoals of abortion politics. BEAC quietly closed its doors on September 29, 1989, having never issued a report. Congress delegated to both commissions the task of using the tools of bioethics to fix problems Congress itself felt powerless to solve. The National Commission became a resounding success; BEAC a political footnote. Some factors influencing the differences can be discerned from reviewing the history of national bioethics commissions and review bodies.

As bioethics continues to expand rapidly as an academic field, its proper role in helping to craft public policy must be re-examined. Bioethics as an academic pursuit grew immensely in the 1970s and 1980s. Students may now obtain degrees in bioethics, at both the graduate and undergraduate level. Thousands have taken courses in bioethics, and hundreds now pursue careers whose core they would describe as bioethics. Ethics committees have proliferated in hospitals and other healthcare institutions, and every research center must have at least one Institutional Review Board. Terms used to erect the framework of bioethical principles over the past twenty years – most notably the "Georgetown mantra" of autonomy, beneficence, and justice – are now heard on hospital wards, in the Halls of Congress, and executive agencies at the State and Federal level. Yet Congress spurned BEAC, a child of its own creation intended to mediate a debate about ethical issues in research and the delivery of healthcare.

Bioethics must find its voice again. The concepts articulated in the Belmont Report of 1978 [24] helped resolve policy debates about the proper way to protect human subjects of research. The principles were translated into regulations (Code of Federal Regulations, 45 CFR 46), and incorporated into daily research practice. But a new, more divisive, and deeper set of issues now confronts the nation, and confounds its political leaders. If bioethics is to help resolve the pressing problems of the day, it must redefine its purposes, devise novel conceptual approaches, and find new methods and mechanisms through which those concepts can be made useful.

The National Commission

The National Commission was created in the National Research Act of

1974 (Public Law 93-348). The bill was sponsored by Senator Edward Kennedy, following a congressional debate dating back to 1968 ([45], pp. 168-189). An initial concern about heart transplantation and the onslaught of new and powerful biomedical technologies was further fueled by news of research abuse. A series of scandals indicated to Congress that biomedical researchers could not keep their own house in order. News of the Tuskeegee syphilis trials, the Willowbrook hepatitis experiments, use of prisoners to test drugs, whole body radiation experiments sponsored by the Department of Defense, and testing of hormone analogs among welfare mothers and Mexican American women were sufficient to overcome investigators and clinicians who resisted what they viewed as interloping from groups outside the profession.

Congress created the National Commission as part of the Department of Health, Education, and Welfare (now the Department of Health and Human Services) and gave it a specific task – to articulate the principles of ethics needed to deal with human subjects in research, and to use those principles to recommend actions by the federal government. The National Commission operated from 1974 until 1978, issuing 10 reports ([19,20,21,22,23,24,25,26,27,28]). Many of these reports were translated, often quite directly, into the now-familiar federal regulations for research involving human subjects (45 CFR 46). The National Commission articulated three basic principles: respect for persons, beneficence, and justice. It laid great emphasis on autonomy, elaborated and extended the notion of informed consent, recognized the special vulnerability of specific populations (e.g., children, prisoners, the mentally infirm), and fleshed out details of review through Institutional Review Boards now embodied in federal research regulations.

The National Commission also had its warts. Paul Ramsey, a close observer, felt that the deck was stacked in favor of research, particularly during deliberations about fetal research [43]. Ramsey nonetheless believed that the fetal research report was one of the best the commission did, and strongly supported the work of the commission overall.[1] The psychosurgery report covered a highly controversial topic, and the Commission's recommendations were largely ignored.[2] Commissioners and staff agreed that the report on research advances had little impact.[3] But the commission clearly left its mark and substantially met its primary mandate.

Congress and the executive branch both found the National Commission useful. The Commission recommended that a successor body be created, but with broader authority to address issues beyond human subjects protection. Congress was again confronted with issues created by the march of new biological and medical technologies. The recombinant DNA debate was in full flower on Capitol Hill, and termination of treatment was rapidly becoming a national issue in the wake of the Karen Ann Quinlan case and other court challenges to medical authority. The debate about a coherent national health plan was on one of its periodic upswings. Congress concurred that a more general mandate was in order, and created the President's Commission for the Study of Ethical Problems in Medicine and Biomedical and Behavioral Research (the President's Commission).

The President's Commission

The President's Commission was established by Section III of Public Law 95-622. Congress specified several tasks, as it had for the National Commission, but also gave it authority to undertake studies at the request of the President or upon its own initiative. The scope of the commission was broadened to the entire federal government. The Commission was an autonomous entity under the President, rather than part of a particular department.

The President's Commission got down to work after being sworn in at a White House ceremony in January 1980. It operated until March 1983 and issued 11 reports, including a summary ([32–42]). The President's Commission extended the human subjects theme of the National Commission in several reports on health research regulations and compensation for research injuries ([33,34,39]). Its report *Defining Death* became the foundation for statutory changes adopted throughout the States [32]. In this report, the President's Commission helped to formulate and explain the Uniform Definition of Death Act. The President's Commission addressed "whistleblowing" in biomedical research [37]. Inaction on this report, and inattention to the related topics of scientific misconduct and fraud, permitted these issues to fester for many years, by which time they emerged as serious embarrassments for the research enterprise. The President's Commission also addressed genetics in prescient reports on genetic screening and counseling [40] and on human gene therapy [36]. The gene therapy report served to ground a potentially explosive discussion and to thwart some ill-considered legislation. The genetic screening report correctly identified issues, but languished amidst federal inaction until the discovery of the gene for cystic fibrosis eight years later rekindled a national debate. The President's Commission reports in these areas were authoritative. They were respected among scholars and useful to policy makers.

The President's Commission also confronted controversies about termination of treatment in its reports on making healthcare decisions [35], and even more directly in *Deciding to Forego Life-Sustaining Treatment* [38]. This report addressed highly charged issues then, as now, much the subject of court decisions. While this report took the Commission into dangerous waters, it illustrated the power and usefulness of a federal bioethics commission. The commission directly confronted the arguments for and against use of life-sustaining treatments. Its emphasis on patient and family-centered decisionmaking captured the prevailing principles of biomedical ethics. The recommendations regarding appointment of surrogate decision makers (e.g., durable powers of attorney) pushed the debate beyond advance directives such as "living wills."

The President's Commission conclusion that nutrition and hydration were not fundamentally different from other medical treatments was perhaps its most controversial finding. This report was issued two years into the Reagan administration, by which time the commission included eight Reagan appointees, in general more conservative than the commissioners they replaced. The report still commanded unanimity among commissioners. The report nonetheless aroused the ire of Senate conservatives who averred that unanimity on the committee masked deep social divisions. *Decisions to Forego* was later cited as the reason that a government body similar to the President's Commission could not be trusted. As Douglas Johnson, registered lobbyist for the National Right to Life Committee (NRLC), put it, the President's Commission "represented itself as stating a consensus when in fact none existed".[4]

The President's Commission explicit goal of unanimous agreement does not imply that other views, including those espoused by the NRLC, were ignored, but rather that the commissioners came to hold positions that diverged from those staked out by interest groups. The legitimacy of the President's Commission was questioned when it encroached on those positions. This political phenomenon will persist in areas of long-standing public controversy [15].

The President's Commission demonstrated that a process that successfully develops consensus among members that represent a full range of initial views can still fail to moderate a divisive debate. This may not be the fault of the President's Commission but an irremediable fact about the political environment. If commissioners are taken to be truly representative, when they change their views in response to facts or arguments that come before the commission, it may be presumed that like-minded Americans with similar starting positions would do the same. But the decisions of individual commissioners may not alter the positions of interest groups that increasingly dominate national debate on controversial topics. Or commissioners answerable to such interest groups may refrain from expressing their individual views to honor outside commitments, thus precluding a consensus that would otherwise form among hypothetical "reasonable persons." Consensus may not be the best goal if a commission is surrounded by powerful outside interest groups, but if common ground cannot be found, then the utility of a national commission process may be greatly reduced for policy makers.

The only means for policy makers to look beyond the purely political nature of such disputes is to assess data and judge the rational arguments presented by different groups. Commissions function well when they take polar positions into account, but if national bioethics commissions cannot expect to satisfy all the strident public interest groups of many persuasions it is unclear how they help. They can forge a consensus among moderate forces, but this is generally not decisive in political battles. The passion generally flows from the poles, where it is transformed into money, time, and organization. For some issues, there is little common ground to stand on, as is the case now for issues related to abortion and withdrawal of fluid and nutrition from moribund patients. There may be little point in having a national bioethics commission attempt to deal with such topics, because commissions are useful principally in generating broad consensus or finding compromise positions that, while not fully acceptable to the opposing factions, are nonetheless endorsed by them because they contain concessions from opposing factions. For groups with narrow interests that answer to a well defined constituency and that brook little dissent, concessions are difficult to accept and even more difficult to offer.

Another President's Commission Report, *Securing Access to Health Care*, was an audacious foray into mainstream health policy, addressing a central issue of American health policy. Staff working on the report were exceptionally able, and the background papers in the appendices are impressive in their diversity and quality. Many of those who wrote for the President's Commission went on to elaborate their ideas in books and scholarly publications. The report on access, however, became the only report that drew a dissenting vote from a commissioner, and this report has been held out for more criticism than most others ([2,4,9], p. 81-83;[10]). The criticisms are well taken, although Bayer and Arras fail to appreciate the political dynamics of the Commission's function. Daniels levels a criticism at the philosophical underpinnings of the report, specifically its diversity of justifications for why healthcare differs from other economic goods and services, that if taken seriously would weaken rather than strengthen the report. He would justify healthcare by its ability to restore a full range of opportunities, but this unfortunately fails to capture some critical areas of care that are not intended to restore lost function, such as long-term care for Alzheimer's disease.

One problem with the access report is that it laid out only the faintest outline of a road map to move from its theoretical base to practical implementation. The principle reason for this is no doubt the fact that the Commission took its mission to be ethical analysis of a range of acceptable policies rather than policy implementation. The dominant issue in healthcare reform was not whether the care system was fair or not, which the Commission could address, but instead was about how to fix the system, a task that any bioethics commission is only questionably competent to address. The report inevitably fell short of a call to arms for a specific program. The report also played little or no role in the most significant health policy change of its day, namely, the adoption of prospective payment by the Medicare program. The President's Commission report remains, however, the only document that explicitly discusses the ethical underpinnings of government intervention to secure access. This attention to justification distinguishes the President's Commission from the Pepper Commission, which laid out a program to attain programmatic objectives, but did not articulate the ethical underpinnings of those objectives (or how to pay for their attainment) [5]. The President's Commission report may yet be viewed as a pivotal event in turning the debate away from a simplistic and unconstrained right to healthcare and towards specification of decent minimum of healthcare. The ultimate utility of the President's Commission report cannot yet be judged, despite a decade of incubation. Whether the access report was far ahead of its time or simply too arcane and theoretical for the policy making community can only be judged as the decades pass.

When it ventured into healthcare financing, the President's Commission entered territory marked by disagreement about political philosophy.

The trees were already marked by ideologies long before the President's Commission wandered into the forest. There is no adequate theory of distributive justice applied to healthcare, although the President's Commission did seek papers from many of the best experts on justice in healthcare.

The National Commission could point to agreement about the immorality of specific research abuses, and could draw on a thirty-year history of legal and public policy debate when it began its work on human subjects protections. The President's Commission could tap general agreement that the healthcare system needed reform, as polls consistently showed consensus that it was unfair, but no dominant framework of how to direct that reform emerged from ethical analysis. The President's Commission could focus attention on unfairness, but could not articulate the solution, since this was largely a matter of economic interests and political will.

The President's Commission here advanced cautiously into political territory heavily patrolled by legions of healthcare professionals, provider institutions, and interest groups. By contrast, the National Commission was by far the most conspicuous body dealing with human subject protections. The President's Commission was but one of many voices speaking about access to healthcare. Many institutions, including several federal bodies, academic centers, and entire foundations, were dedicated to health policy analysis. These analysts had assembled a massive literature on the subject from medical, economic, and public health perspectives. The bioethics literature was sparse by comparison. Bioethics could sketch out the justifications for a healthcare entitlement, but not how to honor it.

Had the Commission continued to exist beyond 1983, it might have been drawn further into the political mainstream concerning health policy issues. The President and Congress might have requested reports on long-term care, coverage of services for the uninsured, maternal and child healthcare, drug treatment, mental health services, and other topics still largely unresolved a decade later. The President's Commission could perhaps have established the same solid track record, with accolades from scholars as well as policy makers, as it had already done in other areas in work, by building on the theoretical base of *Securing Access to Health Care*. If so, it might have have foreshortened the debate about access to healthcare that still persists without substantial policy progress. Instead, the President's Commission passed out of existence.

As the statutory authorization for President's Commission neared its deadline in late 1982, the Commission sought an extension. The Commission also recommended that if it were terminated, a similar body be created. The Commission succeeded in getting a three-month extension until 31 March 1983, but then passed out of existence in a flurry of publications. On balance, the President's Commission left an impressive legacy. Many of its reports had immediate impact on policy, others were sleepers that anticipated issues that emerged later in the 1980s, and most of its reports are still widely cited by scholars in bioethics and in related public policy debates.

The increased complexity of the President's Commission's mandate, compared to the National Commission's, and the increasing presence of interest group politics foreshadowed an impending storm. The two year debate about re-establishing the President's Commission began in November of 1982, when the President's Commission report *Splicing Life* was released at a hearing before Albert Gore, Jr., then a member of the House of Representatives. The hearing focused on the implications of human genetics, particularly gene therapy.

The report emphasized the distinction between genetically altering somatic cells, which would not lead to inherited changes, and germ cells (sperm, egg cells, and their precursors), which would induce inherited changes. This distinction permitted policy makers and others to understand that there were cases of gene therapy that would not be morally different from any other treatment, clearly pointing to some cases where gene therapy might be technically preferable – and morally equivalent – to other treatments [36]. The report steered the debate away from hopelessly vague speculations about playing God and the technological imperative and towards prudent policies of research protocol review.

Splicing Life recommended that the National Institutes of Health review progress in gene therapy through its Recombinant DNA Advisory Committee, and that NIH consider the broad implications of commencing gene therapy. The Recombinant DNA Advisory Committee accepted this recommendation in April 1983 and began to debate the merits of the new technology and to assess its social implications. A working group on human gene therapy was established later that year. The working group proceeded to draft the "Points to Consider in the Design and Submission of Human Somatic Cell Gene Therapy Protocols," adopted in 1986 as the keystone document in public oversight of the new technology ([14,18]). The working group was later reconstituted as the Human Gene Therapy

Subcommittee. When gene therapy protocol review became the main main work of the entire Recombinant DNA Advisory Committee, the subcommittee was dissolved.

Splicing Life also noted that there was a need for public debate, which could be mediated by an ad hoc commission on genetics or by a standing federal bioethics commission. Rep. Gore was impressed with the report, by the process that produced it, and by Alexander Capron, the Executive Director of the President's Commission. Rep. Gore subsequently introduced legislation to create a President's Commission on human genetic engineering, favoring permanent oversight of advances in human genetics and reproduction. This became the seed for legislation to create the Biomedical Ethics Board and Advisory Committee, with a broader mandate than human genetics, as Gore became convinced that a broader mandate would be more useful.[5]

During this same period in the Senate, there were several separate proposals to re-establish the President's Commission, to give the Institute of Medicine (IOM) a mandate to do studies in bioethics, and to have the congressional Office of Technology Assessment (OTA) do such work. A continuation of the President's Commission was unacceptable to several conservative Senators, most notably Gordon Humphrey and Jeremiah Denton who took their cues from the National Right to Life Committee. Their principal objections were grounded in revulsion at the Commission's recommendations about termination of treatment, especially about withdrawal of nutrition and hydration. Senate conservatives wanted bioethics brought under direct congressional scrutiny. Senators Kennedy, Hatch, and Denton sent informal inquiries to the Institute of Medicine and the Office of Technology Assessment to find out if either institution could do the work, and if so whether they were willing to do so. The response of each was equivocal. Some staff at both institutions wanted to move into bioethics, but upper management at both IOM and OTA sensed the political perils.

When the NIH authorization bill went to its House-Senate conference committee in October 1984, Rep. Gore's commission, supported by Reps. Henry Waxman and John Dingell, was proposed in the House version, while OTA was given the task in the Senate version, sponsored by Senators Hatch and Kennedy. Members of OTA's congressional board (John Dingell, Orrin Hatch, and Edward Kennedy) were among the conferees, and they were determined to protect OTA from any unnecessary political risks.

The conference committee created an entirely new agency, the Biomedical Ethics Board and Biomedical Ethics Advisory Committee, structured much like OTA. Anthony Robbins, staff to John Dingell, and David Sundwall, staff to Senator Hatch, took principal responsibility for crafting the compromise, with strong input from staff to Representative Waxman and Senators Humphrey and Denton. The bill was passed by both Houses but vetoed by President Reagan, for reasons unrelated to the bioethics provisions. It was re-introduced in the 99th Congress, vetoed again, but then passed in a veto override by strong House and Senate majorities. It became law as the Health Research Extension Act in May 1985 (Public Law 99-158).

The Biomedical Ethics Advisory Committee

The Biomedical Ethics Advisory Committee (BEAC) was a 14-member group whose multidisciplinary membership was slightly modified from the original Gore bill. The Committee members were appointed by the Biomedical Ethics Board (BEB), comprised of 12 Members of Congress – three Democrats and Republicans from each the House and Senate, similar to OTA's congressional board. It took almost a year for party leaders of the House and Senate to appoint the 12 members of BEB.

The congressional Board appointed the 14 members of BEAC, the operational arm, but only with immense difficulty and after almost 2 1/2 years had passed. Members were to come from three expert categories (roughly: law and ethics, biomedical research, and clinical care) with an additional two lay members. The Board's failure to agree on an appointment process brought IOM and OTA into the picture again, as each vetted nominations submitted by congressional board members. Following OTA and IOM review, the appointment of the 12 experts was completed 22 months after the statute became law. It took another year to appoint the two lay representatives, an exercise that deepened the mistrust that had been building up for several years among members of the congressional Board.

BEAC finally met on September 26 and 27, 1988, less than a week before its legislative authority expired. The committee adopted rules of operation. Alexander Capron was elected chairman and Edmund Pellegrino vice chairman. The meeting went smoothly, despite some confusion about the implications of a vacancy left by the death of appointee Dennis Horan. BEAC members assumed that a new member would be appointed by the congressional board before the next meeting. The second meeting brought experts on several topics in human genetics to speak. The meeting was congenial and productive, yielding a tentative working plan. Indeed, BEAC functioned effectively from the first, in stark contrast to its dysfunctional congressional Board. Even as BEAC began to take flight, the congressional Board to which it was tethered sank deeper and deeper into the political tar pit surrounding abortion.

The first mandated report, on implications of human genetic engineering, stemmed from the original Gore bill proposing an extension of the President's Commission (H.R. 98-2788). The deadline for the second report, on fetal research, expired before BEAC members had been appointed. The fetal research mandate was reinstated in the Omnibus Health Extension Act of 1988 (Public Law 100-607). The third mandate stemmed from Senator Armstrong's proposed amendment to the 1988 AIDS bill (Congressional Record, April 28, 1988, pp. S5009-5022). Following a long and heated debate, the Senate passed an alternative amendment directing BEAC to report on the topic.

BEAC thus had a full plate. But the structure, never stable, began to disintegrate on 8 March 1989, when the Senate BEB members were unable to elect a chairman. The Senators found themselves in a deadlock along partisan and prochoice-prolife lines. The meeting took place at the peak of an acrimonious Senate debate to confirm former Senator John Tower as Secretary of Defense. Partisan tensions were strong, tempers short, and timing inauspicious.

As the congressional Board sank, it dragged BEAC with it. Eventually, the 1990 budget doomed BEAC, since it could not use its appropriation until its parent Board accomplished tasks that had become impossible to achieve. The appropriations rider happened to be introduced by BEB member, Senator Don Nickles, a prolife conservative, but Rep. Waxman, a prochoice liberal, was equally vexed with the enterprise, and expected to take similar action if conservatives failed to do so. BEAC died in the cross-fire. The conditions stipulated in the appropriations bill were not met before the new fiscal year began, and the office in the Hart Senate Office Building was closed at the end of September, 1989. BEAC was not alone in foundering in a difficult political environment. Two other federal bodies encountered similar problems during this period.

The Ethics Advisory Board and Fetal Tissue Transplantation Research Panel

While BEAC brought bioethics to Congress in late 1988, the Department of Health and Human Services took several actions that promised to bring bioethics back into the Executive Branch. An Ethics Advisory Board (EAB) had existed from 1978 until 1980, in the interregnum between the National and President's Commissions. The EAB issued a 1979 report on in vitro fertilization, stipulating several criteria for approval of such experiments [12]. The EAB recommended granting a waiver to permit fetoscopy to diagnose hemoglobinopathies, and handled several issues related to Freedom of Information Act inquiries in the Department. Joseph Califano, Secretary of Health and Human Services, granted a waiver permitting a particular research project to go forward, although it was moot as the principal investigator died. Secretary Califano was soon replaced by Patricia Harris, and the general policy recommendations of the EAB were never addressed. The budget for the EAB was diverted to the President's Commission in 1979. Most funding for the President's Commission was derived from other sources, but the link between dissolution of the EAB and creation of the President's Commission betrayed a confusion on the part of policy makers about their distinct purposes.[6] The EAB was positioned to review protocols that raised novel issues and to devise procedures and criteria for executive review; the President's Commission was a forum for national debate.

Reproductive issues forced debate about the EAB to the surface again in 1988, when OTA released a report on treatment of infertility [56]. A subsequent House hearing before Congressman Weiss in June 1988 focused on the absence of the EAB, in violation of the Department's own regulations. Assistant Secretary of Health Robert Windom promised to re-establish it. The proposal to charter a new EAB was published in the September 12, 1988, *Federal Register*, with the approval of Secretary Otis Bowen (Ethics Advisory Board; Notice of Establishment *Federal Register* 53 (No. 176): 35232-3). The comment period passed. A revised charter was drafted but never signed in the waning months of 1988, following the election of President Bush. The new administration did not support reestablishing an EAB.

Another foray into national bioethics began in May 1988, when Dr. Windom initiated a moratorium on the use of fetal tissue in transplantation research funded by the federal government. He requested that NIH convene a panel to advise him about the technical stakes and ethical implications of such research. NIH complied, and scheduled a meeting for September 1988. These events followed many months of wrangling among NIH, the HHS Secretary's and Assistant Secretary's offices, and the White House. The initial, unrealistic expectation was that an ad hoc advisory panel would make recommendations after a single meeting. The issues predictably proved far too complex and divisive. The panel met to hear testimony from disease groups, researchers, and those opposed to the research. They voted on a set of specific recommendations, and found a majority in favor of permitting such research as long as three conditions were met in addition to IRB approval: 1) the decision to donate tissue was kept separate from and made only *after* the decision to abort, 2) the process for abortion was not altered in any way, and 3) the informed consent of both parents was obtained in cases when the father could be contacted. The majority of the panel argued that they did not have directly to engage questions about the morality of abortion, since the practice was legal [6].

Two separate dissenting statements argued that the abortion issue should not be sidestepped. One statement was introduced just as the panel was preparing to adjourn. It was written by James Bopp and James Burtchaell, who dissented from the majority on three grounds: 1) that the process of using fetal tissue was inherently suspect because it would give women an altruistic incentive to perform an immoral act (i.e., elective abortion); 2) that such research fostered complicity in the practice of elective abortion because it would necessarily entail close collaboration with those performing abortions; and 3) that permission from the mother and father were illegitimate, since parents abandoned the fetus by seeking the abortion.

The panel met twice more, in November and December 1988. A report was issued with a compilation of recommendations and vote tallies, accompanied by two dissenting statements, by Bopp and Burtchaell and by Rabbi David Bleich, several separate assenting statements were signed by subgroups of the majority, taking issue with points made in the dissenting statements [7]. The panel's report was then approved by the advisory committee to the director of NIH in December 1988, urging acceptance of the recommendations [1]. The panel's majority recommendations were ultimately rejected by Assistant Secretary of Health James Mason, in favor of the arguments of the prolife minority.

The Office of Technology Assessment

The Office of Technology Assessment (OTA) was established in 1972 as an analytical arm of Congress, to anticipate how science and technology would raise issues for policy makers, and to advise Congress on federal policies affecting science and technology development. It was to render technical advice about how to promote or regulate science and technology, and to give early warning about the impacts of emerging technologies. In a 1983 report on Genetic Testing in the Workplace [47], OTA began to explicitly incorporate bioethics into some reports. A 1984 report on Human Genome Therapy followed in the steps of the President's Commission [48]. A succession of reports included chapters on ethical considerations, or had extensive discussion of ethical issues ([47–51,54–60]). Two OTA projects jointly commissioned a set of bioethics papers subsequently published as a 1986 supplement to the *Milbank Quarterly* ([52,53,65]). Many other reports on biomedical research and health policy built on papers commissioned from academic bioethicists. Clearly, bioethics was becoming a component of the OTA process in several areas, but OTA was not a bioethics commission. Ethical analysis was usually but a component of selected projects.

Bioethics was the focus of an occasional OTA report. Four reports stood out as exemplars of bioethics at OTA. *Human Gene Therapy, Ownership of Human Cells and Tissues, Infertility*, and *Neural Grafting* were all reports on topics that could well have been assigned to a federal bioethics commission had one existed. Thus, while OTA had no explicit mandate to serve as a bioethics commission, its mission of technology assessment extended into mainstream bioethics on several occasions. Indeed, the question of whether and how to structure federal advisory bodies on bioethics became the topic of an OTA meeting in December 1992, which produced an OTA background paper the following year [67]. OTA's potential to contribute to federal bioethics expired when the 104th Congress killed the agency in October 1995.

The Institute of Medicine and National Academy of Sciences

The Institute of Medicine (IOM) was created in 1971, under the charter of the National Academy of Sciences (NAS) of which it is part. The NAS charter was passed as an act of Congress in 1863 and signed by President Abraham Lincoln. IOM and NAS each has its own membership, elected by current members for their contributions to medicine and science, respectively. In addition to this honorific function of conferring prestigious membership on distinguished individuals, both IOM and NAS also have operational arms that conduct policy studies. Some of these studies are focused on topics in bioethics, or contain sections that deal with such topics. Indeed, IOM's early history paralleled the growth of academic bioethics as a discipline, contemporary to the National and President's Commissions. An IOM division on Legal, Ethical, and Educational Aspects of Health operated for several years, but proved difficult to sustain financially. It produced several reports on ethical issues in healthcare, medical practice, and health promotion. Other divisions, particularly the division of Health Science Policy continued work in bioethics when that division died in the late 1970s. In particular, IOM produced two directly relevant reports in the early 1990s. One of these dealt with use of genetic tests [68], and built on an NAS report of two decades earlier. The other, *Society's Choices* systematically surveyed the practice of bioethics and its relation to policy at various levels, and provided detail far beyond that covered in this chapter [69].

Genome Ethics and Other New Federal Initiatives

While several federal bioethics bodies came a cropper amidst abortion politics, other parts of the federal government made new commitments to bioethics. The National Center for Human Genome Research at NIH and a parallel program under the Department of Energy announced their intention to devote roughly three percent of their genome research budgets to analysis of the social and ethical implications [63]. With a total joint budget of almost $86 million for 1990, this constituted over $1.6 million, more than trebling the federal contribution to bioethics research [8]. The NIH fraction escalated to almost 5 percent in 1991, or $4 million, with DOE contributing another $1.4 million. This increased the federal commitment to bioethics severalfold again, and future increases in the total genome research budget continued the trend.

Earmarked funding for bioethics research, broadly construed, was a noteworthy policy innovation. It placed such research on a similar footing with other research. The fact that only genome research programs made such commitments, however, changed the face of bioethics. The immense increase in federal sponsorship was restricted to human genetics alone, spawning a more specialized breed of bioethical analyst. The National

Center for Nursing Research at NIH began a much smaller grants program on ethical issues related to nursing care, restoring the balance to some extent, but until other institutes at NIH followed suit, the bioethics agenda would be skewed heavily towards genetics.

Bioethics programs under the genome research flag were the most conspicuous and substantial in the federal government, but they were not entirely unprecedented. The Ethics and Values Studies (EVS) Program of the National Science Foundation long supported analysis of ethical issues in areas related to biology. The NSF Program tended to avoid issues related to clinical care, since its mandate was to inspect issues related to science and technology rather than medical practice, but the EVS program did support more than a dozen bioethics projects over the years, including analysis of human gene therapy, recombinant DNA policy, and other topics.

The National Endowment for the Humanities also supported several projects in bioethics, typically courses, book projects, or workshops. The NEH program's goal is to improve the quality and impact of the humanities per se, not the analysis of public policy. This contrasted strongly with genome funds intended to illuminate national policy options.

The Ethical, Legal and Social Issues (ELSI) program spawned by the human genome project was not only a grant-making function similar to other research, but also a policy-oriented governing body. The idea for a Human Genome Initiative was first hatched by Charles DeLisi in the Department of Energy (DOE) in late 1985 ([8,11]). In 1986, Dr. DeLisi began informal talks with bioethicists at the Kennedy Institute of Ethics at Georgetown University, but he left government service before plans reached fruition. The National Research Council redefined the scope and purpose of the Human Genome Project in a report issued in February 1988 [29]. The congressional Office of Technology Assessment corroborated the findings of that report three months later, although suggesting a somewhat different administrative structure [57]. Both reports drew attention to the social, legal, and ethical implications of advances in human genetics. The rapid advances in technology and knowledge would lead to refined diagnosis, genetic testing and genetic screening. New technical capacities and new knowledge would, in turn, change genetic counseling, and would exacerbate existing problems surrounding confidentiality and privacy. Most troubling was the prospect of genetic discrimination – for example, unwarranted constraints placed on those seeking jobs, insurance or other social benefits by dint of their genes.

James Dewey Watson announced his intention to establish a research program on ethical, legal, and social issues when he accepted the mantle as director of NIH's Office of Human Genome Research in September 1988 [62]. Dr. Watson made this commitment on his own. It came as a surprise to NIH staff, including NIH Director James Wyngaarden. In subsequent discussions with reporters, this general commitment homed in on the specific figure of 3 percent of the total budget [46]. A Working Group on Ethical, Legal and Social Issues of human genome research was appointed under the NIH's genome research advisory committee. The ELSI Working Group was chaired by psychologist Nancy Sabin Wexler from Columbia University, best known for her advocacy of research into Huntington disease and her concern for adequate safeguards on genetic testing for that condition.

NIH's Office of Human Genome Research gained budget authority a year later, becoming the National Center for Human Genome Research through administrative action by HHS Secretary Louis Sullivan, Jr., at Dr. Wyngaarden's suggestion. By then, the 3 percent figure had become a floor, not a ceiling. At Senate hearings in November 1989, Albert Gore, now a Senator, praised the NIH for its ELSI program, and excoriated DOE for failing to make a similar budget commitment. A month later, an external advisory committee reporting to both NIH and DOE recommended adopting the ELSI Working Group as a joint effort. The ELSI Working Group had an informal role in overseeing the research portfolio of NIH and DOE, and helped formulate the requests for grant proposals and program announcements. Its mandate went beyond this to include responsibility to formulate policy options. The importance of this function was reinforced by the Senate Appropriations Committee in hearings for the fiscal year 1991 budget [61].

The ELSI working group operated as an advisor to the joint NIH-DOE genome research committee, and made policy statements on the need for pilot studies of cystic fibrosis screening [66], and on protection from genetic discrimination under the Americans with Disabilities Act [16]. These forays into policy activism touched on human genome research. The Working Group could not, and did not aspire, to replace the functions of a national bioethics commission.

II. WHAT IS THE PURPOSE OF BIOETHICS IN PUBLIC POLICY?

The attraction of bioethics for policy makers grew from success in solving problems through a broad and systematic interdisciplinary process. The National Commission was useful; later bioethics commissions were intended to replicate its success. The National Commission's strategy of "bootstrapping," alternating between theoretical principles of ethics and practical analysis of concrete situations, became a powerful method [13]. The discussions were translated into recommendations and implemented through regulations. The National Commission solved the problems related to use of human subjects in research better than Congress or the executive branch expected. The National Commission laid a foundation for policies that could have been long delayed or buried without the Commission's contribution, although the historical verdict is blurred by the fact that NIH was already progressing in the same direction before the National Commission was founded.[7]

The recommendation of a national commission is much more than another book by an academic group, and generally has more clout than a work of individual scholarship. The various documents of the national independent bioethics commissions are widely used in courts to decide actual cases, in federal and state legislatures to devise statutes, and are also intellectual and policy landmarks. They are regarded as the best efforts of the nation's thinkers. National commissions present mechanisms to deal with issues amenable to a search for consensus.

The thinking and writing processes are not the only ones that take place, and in some cases these are not even the most important. Instead, the process of writing a report for the federal government can be an opportunity to create the environment in which political action becomes possible, and to inject relevant information into the process of formulating national policies. The process of gathering facts and assembling an advisory panel can encourage coalition-building among parties previously unknown to or suspicious of one another. The same process used to ensure balance and accuracy can forge new alliances among the parties, and the process of consensus building requires that those whose interests are at stake clearly articulate their concerns and listen to the concerns of others. Managed sagaciously, this process can have long-term benefits well beyond production of a report.

Writing for the federal government brings access to the expertise of the nation, and the power to convene highly respected representatives of disparate groups. The analysis goes forward between major centers of political power. For any given activity related to medicine or biology, there are several hundred people making decisions in the federal government. There will typically be ten to twenty "key" individuals – usually budget gatekeepers, rule-makers, or nodes in the flow of information. In healthcare, this includes members and staff of the tax policy committees, the health committees, and the appropriations committees in Congress. Health policy in the Executive Branch involves the Health Care Financing Administration, Agency for Health Care Policy and Research, and other parts of the Department of Health and Human Services, the Department of Veterans Affairs, the Department of Defence, and other executive agencies. For biological science and biomedical research, there are committees that authorize and appropriate funds for the agencies within the Department of Veterans Affairs, the Department of Health and Human Services, the National Science Foundation, the Department of Energy, and other agencies conducting biological research. There are also contacts at each agency, the Office of Management and Budget, and the Office of Science and Technology Policy. Power centers in the United States are decentralized to the degree that failure to gain support, or at least acquiescence, from any of these key people can doom any proposed change.

A focus on key decisionmaking positions, and the individuals who occupy them, is an essential feature of successful public policy analysis. Cultivation of the interest and respect of those who must devise and execute policies is therefore necessary. One must solicit the views of experts knowledgeable about the topic and simultaneously learn about who will be making crucial policy decisions. Information is gathered from as many sources as possible, using papers commissioned from experts, interviews, and meetings. The next step is to filter the information, to identify the people whose judgment is unclouded by their personal passions (always a matter of degree), and to construct the main lines of argument. This process will generally reveal areas of consensus and areas where there is none. Arguments are then articulated in preliminary documents that are then distributed to at least two key constituencies: the experts and the decision makers. This step brings the groups together, and makes the report-writing team a focus for information exchange. If successful, the team garners the esteem of both experts and policy makers, which facilitates a faster and more open flow of information – from the experts about their judgments and from policy makers about goals and concerns.

A truly successful national report is not merely a book or pamphlet, but also a process to cultivate mutual respect among the communities involved. Kenneth Arrow describes the value of trust and the costs of mistrust in *The Limits of Organization* [3]. Writing a report, when it goes well, creates an environment for such trust to develop among the various communities involved in policy formation. It is, needless to say, a fragile process and demands relentless attention. In a typical project, fully half of the effort may be devoted informally to conveying technical information to decision makers and information about the practice of government to technical experts. A report is often the most conspicuous output, especially to scholars, but the underlying process is often more important in influencing policy change.

One of the toughest issues in writing national reports is how to make controversy productive. There is always controversy related to any issue about which a national body is asked to deliberate (or there would be no reason to investigate it). Controversy is useful when it can be channeled into responsible positions, and articulated as competing viewpoints of equal, or at least mutually respecting, merit. This is a crucial function of national committees. It means, however, that national committees are not usually laying the intellectual foundations for ethical analysis. The process is more analogous to sculpture from "found objects" than carving a statue of David from virgin marble.

National commissions cannot usually traffic in legal, ethical, and technical arcana. Their audience is impatient with hair-splitting and oblivious to philosophical niceties. Richard Momeyer notes that "concern with rarefied notions of truth and the integrity of seekers of truth in the domain of policy formation distracts one from the real point of that process, which is to arrive at decent, workable policies that show due regard for all relevant interests" [17]. There are some conspicuous exceptions to this rule, as any other relating to democratic government. The very small semantic shift from "right" to "social obligation" in the *Securing Access* report of the President's commission is arguably a case in point. Views differ on whether this was a distinction without a difference or whether it successfully braked a runaway "right" to any service or technology that could enhance health, regardless of cost or consequence.

National commissions must focus on identifying different views of the world, and juxtapose those views so that agreement is brought to the surface and disagreement is made explicit. This is not, strictly speaking, bioethics, or at least not all of it. Bioethics must also involve careful moral justifications and sustained arguments. This is, of course, a matter of degree, and if a national commission is wrestling with issues for which the analytical framework has not been constructed, it may find that is *has* to undertake formal analytical justification from first principles. The National Commission's Belmont Report, for example, broke new intellectual ground. The most useful National Commission, President's Commission, OTA, and other national bioethics reports compare favorably with the best of academic bioethics.

More fundamental, philosophical bioethical analysis – such as writing books on general theories of justice – is best done by those whose job is to think and write, not a group of commissioners selected because of their national prominence, professional background, political connections, and ideological affiliations. A national committee is better adapted to articulate others' positions, filter information, and facilitate communication among policy makers.

Reports produced by the National and President's Commissions, as well as OTA and other federal groups, perform a broad range of functions. Some of the major functions are described below.

Crystallize a Consensus

Some public issues arise so quickly that controversy surfaces as a symptom of incomplete analysis by different factions. In such cases, there is an opportunity to articulate a position that commands wide assent. In political terms, this is extremely useful because democratic government is erected on consensus (although constrained by rights). Human gene therapy is an example – consensus that somatic cell gene therapy is little different form other medical technologies was voiced first by the President's Commission and then by OTA. This was sufficient to prevent legislation to prohibit or limit it. The President's Commission report on defining death, and many of the National Commission's reports on human research subjects also exemplify this function of forming the point of condensation for consensus. Consensus formation is ideally suited to national commissions or ad hoc commissions.

Articulate Residual Points of Disagreement and Clarify Values

Consensus does not always form around issues, even when a technology is the main new feature. Although technological advance is a critical element, for example, nuclear power and arms control have not been notable for their rational public discourse or clean and highly analytical policy process. In bioethics commissions, the objective may be consensus but the result may be incompatible views. The process of articulating this is nonetheless valuable. Seeking consensus may also contribute to public debate with no expectation that concrete recommendations are possible. Thus consensus-seeking need not be considered a failure if it yields progress but no end. There is ample value in ventilating disparate moral positions publicly. An illusory or forced consensus can result in policy change only to breed later backlash as the policy encounters resistance.

Identify Emerging Issues

National commissions can identify future issues. This can be quite helpful to policy makers even if no conclusions are reached about options for dealing with the issues. "Early warning" functions constitute specific, narrow, example of consensus formation. The focus is on identifying issues likely to matter in the future rather than on solutions to current policy concerns. Policy makers value outside advice on future issues because politics is, in part, a process of hearing about parades in time to be at the front instead of the back. If a set of issues is agreed to be important in the future, then a politician can begin to formulate his or her positions on those issues, and to commit resources to finding out about them.

Serve as Forum for National Debate

If an issue is widely agreed to elude consensus, national government may nonetheless benefit from a debate that pits the best minds of various camps against one another in a mutually respecting forum. More often, however, little value will be added by the commission's deliberations. Fetal research is an example of an issue that has been addressed by the National Commission, then an Ethics Advisory Board, and was to be addressed by BEAC. No action to break up the ideological logjam has been effective.

The value of a forum for national debate, if it does not produce consensus, depends on a complex judgment of what the alternatives are, and the probability of advancing the debate by focusing it. Even if consensus is not possible, a softening of positions at the extreme edges may be. If this occurs, the environment for making tough policy choices may be less threatening, and incremental policy adjustments may become possible.

Some issues bound to elude consensus are relatively easy to identify. They usually have been hotly debated for years (e.g., human rights, fetal research, abortion, access to healthcare, surrogate motherhood, homosexuality). For such issues, the bulk of work must be done in the academic community because national commissions are powerless to do more than articulate old disagreements. Changes on such issues require either extended careful thought, followed by changes in public attitudes, or classical political maneuvering to which ethical analysis contributes little. If continuing a controversial debate is the objective, there is generally little to be gained by having a national commission. Some reports fall into this category because a topic thought ripe for consensus proves not to be (for example, the President's Commission report *Compensating Research Injury* ([34,64]).

Excuse Delay and Ground Controversy

Policy makers often hope that controversy will subside. They may seek to use a "study" as a delaying tactic, judging that the intensity of conflict will dissipate over the course of a mandated study. A closely related tactic is for policy makers to call for a study, while politically maneuvering to distance themselves from its results. When the results are produced, they can accept the results they agree with, and blame the commission for those they reject. The BEAC mandates on fetal research and on nutrition and hydration of dying patients were both examples of this strategy. Prolife conservatives mandated the fetal research study to extend a statutory moratorium, while liberals called for a study of nutrition and hydration to frustrate Senator William Armstrong's attempt to amend an AIDS bill. Had these studies been carried out, they would likely have been politically "hot." Having satisfied the need for delay they would next have become lightning rods for acrimony. BEAC was thus not entirely bereft of accomplishments, as it served as a legislative shield for both conservatives and liberals. Both factions must have remembered how BEAC thwarted them more vividly than how it obstructed their adversaries, or they might have kept it alive.

Craft Legislation or Promulgate Regulations

A commission can, having identified an existing consensus, devise a way to incorporate it into practice. The President's Commission report *Defining Death* served this function, as the template for statutes passed in the States [32]. The National Commission reports on children, prisoners, and other vulnerable populations were readily translated into federal regulations governing research. In Congress, this function is usually performed by committees, which have access to outside expertise and focus it on legislation. Executive agencies also have policy-making groups that perform analogous functions. From time to time, however, policy makers may wish to attend explicitly to the ethical dimensions of a policy choice. In such cases, a national commission or ad hoc panel is the logical choice.

Review Implementation of Legislation and Regulation by Executive Agencies

Government has diverse methods to oversee policy. Congress has oversight for the executive agencies, and each agency has its own formal system of checks and balances. The various inspector generals' offices and the congressional General Accounting Office have oversight as their main functions. From time to time, however, the standard process breaks down, is prevented from working, or is inadequate to a particular task. In such cases, a "blue ribbon" commission may be the only politically acceptable alternative. Asking a national commission to review implementation of policy is a symptom that routine oversight has failed. The Tower commission to investigate the Iran-Contra affair and the Warren commission on the Kennedy assassination are examples of these *ad hoc* commissions to investigate specific events. In the bioethics arena, the Human Fetal Transplantation Research Panel is an example ([1,6,7]).

The National Commission was involved in several activities that fall into this category. One impetus for creating the National Commission was, after all, the disclosure of ethically unacceptable practices in research on humans. The National Commission was also asked to review the consent form used for the national swine flu vaccination program beginning at that time. The National Commission review revealed several deficiencies in the form used by the Centers for Disease Control (CDC). CDC was committed to going ahead with the vaccination program quickly, however, and the National Commission's involvement was regarded as unwelcome outside interference. (One can only speculate whether the problems that later plagued the federal government as a consequence of the swine flu program could have been avoided if CDC heeded the commission's advice).

The National Commission reviewed existing federal statutes and regulation for the adequacy of human subjects protection, and scrutinized the processes governing their implementation. The President's Commission followed this precedent by reviewing processes for whistle-blowing in scientific research, and the functioning of Institutional Review Boards. The prominence of the National and President's Commissions was instrumental in their generally successful oversight of executive agencies. The ability to command a response from the agencies within 90 days was essential to enforce accountability. The issues in question were largely those for which there was a sense that agencies were not following optimal policies, and the commissions served as mechanisms to make continued neglect embarrassing.

Aid Judicial Decisions

Many issues in American society are decided in the courts. This is often because legislatures have not found politically viable solutions to problems or have not identified issues in advance. In such cases, the judicial system is left to clean up the mess. The courts are not well equipped to deal with certain questions, however. Judgments requiring extensive practical experience (e.g. medicine) or technical expertise are difficult for the courts to weigh. Judges and juries often must decide on the basis of incomplete and conflicting information. In such cases, the report of a national deliberative body can be useful. Many of the President's Commission reports have been used in this way, cited in court decisions to terminate life support, for example. When this use is anticipated, legal expertise must be incorporated into the analysis. The legal expertise of the President's Commission and its staff was a great strength in this regard.

Create a Critical Mass of Accessible Expertise

All public bioethics commissions must base most of their work on previous analysis. The function of a national report-writing team is usually to articulate borrowed arguments. Linking statements from disparate sources into a single report can nonetheless have revolutionary consequences. But the arguments, by and large, must be manufactured elsewhere. A national commission cannot juggle all the needs of its technical and political clients and at the same time write *A Theory of Justice, Reasons and Persons*, or *Anarchy, State and Utopia* ([30,31,44]). For this, academic research is necessary. Fundamental new work requires years of contemplative thinking and cannot be done by committee. The President's Commission and OTA reports on gene therapy, for example, depended in large part on the writings of those doing the scientific work who had pondered the wider implications of it. They also depended on articles and books written by legal and ethical scholars. Government cannot directly do this sort of academic work, but it can support it through grants and training programs at universities and research centers.

The existence of an analytical infrastructure is crucial to the success of more prominent national commissions. This means that support for such activity must be assured before national commissions can be useful. Yet the role of government in fostering academic analysis is not widely agreed upon. In the United States, applied ethical analysis has prospered despite a dearth of federal support. This is because the academic structure is large and has, until recent years, enjoyed a long period of relative good health (especially in medicine). Universities have taken the lead in directly supporting bioethics research, in contrast to biomedical research in which the federal government and private industry dominate research funding.

The world has changed since 1978, and some topics can no longer be productively engaged by a national bioethics commission. Abortion elicits strongly held but incompatible views that rational people reach from different moral premises. The missing element is not clearly articulated positions that can be filtered and analyzed by a commission, but rather a mechanism to force the cacophony of dissenting voices to deal with the best arguments of the opposing factions, and to agree to practical positions acceptable to the full range, or at least a broad range, of views. By dismantling BEAC and BEB, backing away from the EAB, and failing to implement the NIH fetal tissue recommendations, policy makers indicated that the process of public policy formation was too thankless to be worth the political risks.

Interest groups have become much more sophisticated in their use of national direct mail fund-raising, organization of national letter-writing campaigns, boycotts, and other tactics [15]. They introduce a new dynamic into the political process. They are organized around specific issues, and establish a staff, newsletters, policy analysis mechanisms, and capacities for political strategy that once formed can be applied to new issues as they arise. The great strength of interest groups is their narrow focus, which permits them to concentrate on a specific agenda. But this can also be a weakness, as it tends to result in fixed policy positions that once taken are extremely difficult to modify. A narrow focus can lead to parochial policy formulations; consequences of policy recommendations may not take account of their broader impact outside the sphere of interest. Those representing interest groups may well find it difficult to accept any positions outside the narrow range permitted by their group's policy-making board, as they must answer to a defined constituency. There was substantial controversy surrounding the deliberations of the National and President's Commissions, but those groups did not work in an environment so dominated by public interest groups.

The intrusion of interest group politics into the biomedical area may or may not be lamentable. They have introduced a new level of organization and intensity to political action in many areas. They have also narrowed the range of issues amenable to consensus formation. Issues in the grip of powerful interest groups may now fall beyond the grasp of national bioethics commissions. Any future bioethics commission must either adapt to interest group tactics, or avoid topics that encroach on interest group positions. This applies to any questions directly related to abortion or even indirectly associated with right-to-life questions. The abortion debate is now ramifying into other areas, such as termination of treatment, in vitro fertilization, and fetal research. The divisive debate about use of animals in research seems equally unpromising. Whether the continued growth of grassroots bioethics can eventually produce a more robust consensus in these areas remains to be seen, but for the foreseeable future, bioethics commissions are unlikely to make a substantial contribution here.

The jury is still out regarding access to healthcare. The debate about healthcare financing does not hinge on public revulsion at being mistreated as a research "guinea pig," and thus lacks the motivating agreement about concrete immoral acts that disposed towards success of the National Commission. Many of the lives disrupted by an unfair healthcare system are statistical lives. Stories of fiscal ruin induced by unemployment or inability to secure insurance are wrenching, but lack some of the life and death drama that drives other policies. The Tuskeegee trials, which involved primarily African American males, brought the repugnant issue of racial discrimination into considerations of human subjects protections; the Willowbrook trials among mentally retarded adolescents elicited strong protective emotions. Healthcare financing affects more people, but may seem a less emotional and immediate concern to most Americans, at least until they are personally confronted with a family tragedy or a large hospital bill. Access to healthcare looms large as a major problem for American healthcare, but it is as yet unclear whether bioethicists – as opposed to provider groups, consumer groups, economists, and mainstream health policy analysts – can clearly illuminate a path for policy makers to follow.

In highly technical areas, such as genetics, or in areas hitherto neglected by bioethics, such as nursing care and long-term care, the prospects for a useful national bioethics commission seem bright, if a mechanism to address them can be established. Some issues are typically associated with new practices whose novelty makes it less likely that public interest groups will have taken firm positions. The possibility of third parties, such as private employers or insurers, using genetic tests for purposes of discrimination, for example, raises serious policy dilemmas, but not of a kind that provokes deeply emotional disagreement breaking along lines related to abortion or animal rights. Policy makers have acknowledged the need for a better system of long-term care for over a decade, but the goals of the system and the moral premises upon which it might be based have not been subject to extensive analysis. This would appear to be another area ripe for productive engagement by a national bioethics commission. Policy makers might well benefit from explicit ethical analysis of several other topics. A short list of possible topics that could be usefully addressed by a federal bioethics commission might include:

- Access to long-term care
- Financing healthcare for those without private health insurance;
- Recommendations governing use of human research subjects in nursing homes, home care, and ambulatory settings;
- Recommendations governing use of human research subjects in social science;
- Participation of institutions in organ-sharing for transplantation;
- Confidentiality of results from AIDS tests, drug tests, and genetic tests; and
- Use of medical test results by employers and insurers.

Several of these issues were recognized by previous national bioethics commissions, but not addressed by them because of time or resource limitations, or because they fell outside their mandate. The Bipartisan Commission on Comprehensive Health Care (the Pepper Commission) dealt with the first two topics, and succeeded in reaching some broad conclusions about policy direction [5]. The Pepper Commission thus clarified where health policy might head in the 1990s, but left further tasks for policy makers. The Pepper Commission did not provide either the theoretical foundation for broadening healthcare entitlements or the financing mechanism to pay for the programs it recommended. A bioethics commission is unlikely to help with the details of financing, but might well make an important contribution by examining the foundations of long-term care policy and reexamining the role of government in ensuring a decent minimum of healthcare. This would provide a strong incentive for a more robust theory of communitarian ethics, a notoriously weak area in bioethics and moral theory that is now receiving greater attention. It would also clarify the moral arguments necessary for citizens and their representatives to choose among health policy options, potentially having a direct impact on the momentum for policy change.

A national effort to incorporate ethical analysis into public policy rests on an academic reservoir of technical experts, legal scholars, and humanists. If no critical mass of people in these fields exists, then the first step in a national program must be to develop one. Grants and training programs are the direct means to this end. If there is sufficient expertise in the various fields, then ad hoc committees, national commissions, or permanent analytical agencies are all possibilities. Choosing among these options will depend on the number of issues at hand, the resources available, and the objectives of seeking advice. If consensus is a likely outcome, and publicity is desirable, then a national independent body or ad hoc committee is the logical choice. Care must be taken, however, to provide sufficient budget and time. Funds and schedules must, in particular, allow for the extensive network formation necessary for a proper job. If there are many issues and the decisionmaking apparatus is complex, then a permanent analytical agency is the option of choice. In this case, the extra investment in a management structure is necessary in addition to the report-writing team or teams.

National Academy of Sciences
Washington, D.C., U.S.A.

NOTES

1 According to Ramsey in a lecture at the Kennedy Institute of Ethics, May 1987.

2 Personal communication by Mishkin, Hogan and Hartson, 1983.

3 This was in itself useful, as the Commission concluded that the issues merited further attention and required resources beyond those it could muster.

4 Personal communication, July 1989.

[5] Alexander Morgan Capron, University of Southern California, personal communication, March 1990.

[6] Alexander Capron, personal communication, March 1990.

[7] Personal communications, Barbara Mishkin (October 1989), Hogan & Hartson, Washington, DC; Charles McKay (September 1991) and Charles McCarthy (July 1989), Office of Protection from Research Risks, National Institutes of Health.

BIBLIOGRAPHY

1. Advisory Committee to the Director: 1988, *Human Fetal Tissue Transplantation Research*, National Institutes of Health, Bethesda, MD.
2. Arras, J. D.: 1984, 'Retreat from the Right to Health Care: The President's Commission and Access to Health Care', *Cardozo Law Review* **6** (2), 321-346.
3. Arrow, K.: 1974, *The Limits of Organization*, WW Norton, New York.
4. Bayer, R.: 1984, 'Ethics, Politics, and Access to Health Care: A Critical Analysis of the President's Commission for the Study of Ethical Problems in Medicine and Biomedical and Behavioral Research', *Cardozo Law Review* **6** (2), 303-320.
5. Bipartisan Commission on Comprehensive Health Care: Pepper Commission: 1990, *Recommendations*, Bipartisan Commission on Comprehensive Health Care, US Congress, Washington, DC.
6. Childress, J.: 1991, 'Deliberations of the Human Fetal Tissue Transplantation Research Panel', in K. Hanna (ed.), *Biomedical Politics*, National Academy Press, Washington, DC, pp. 215-248.
7. Consultants to the Advisory Committee to the Director: 1988, *Report of the Human Fetal Tissue Transplantation Research Panel*, National Institutes of Health, Bethesda, MD.
8. Cook-Deegan, R. M.: 1991, *Gene Quest: Science, Politics, and the Human Genome project. A Scholarly Prepublication Draft*, National Reference Center for Bioethics Literature, Georgetown University, Washington, DC.
9. Daniels, N.: 1985, *Just Health Care*, Cambridge University Press, New York.
10. Daniels, N.: 1988, *Am I My Parents Keeper? An Essay on Justice Between the Young and the Old*, Oxford University Press, New York.
11. DeLisi, C.: 1988, 'The Human Genome Project', *American Scientist* 76, 488-493.
12. Ethics Advisory Board, US Department of Health, Education, and Welfare: 1979, *Report and Conclusions: Support of Research Involving Human In Vitro Fertilization and Embryo Transfer*, Government Printing Office, Washington, DC.
13. Faden, R. R. and Beauchamp, T. L.: 1986, *A History and Theory of Informed Consent*, Oxford University Press, New York.
14. Fletcher, J. C.: 1990, 'Evolution of the Ethical Debate about Human Gene Therapy', *Human Gene Therapy* **1** (1), 55-68.
15. Hanna, K. E.: 1991, *Biomedical Politics*, National Academy Press, Washington, DC.
16. Joint Working Group on Ethical, Legal, and Social Issues: 1991, *Genetic Discrimination and the Americans with Disabilities Act*. Submitted to the Equal Employment Opportunity Commission, 29 April.
17. Momeyer, R. W.: 1990, 'Philosophers and the Public Policy Process: Inside, Outside, or Nowhere at All?' *The Journal of Medicine and Philosophy* **15**, 391-409.

18. Murray, T. H.: 1990, 'Human Gene Therapy, the Public, and Public Policy', *Human Gene Therapy* **1** (1): 49-54.
19. National Commission for the Protection of Human Subjects of Biomedical and Behavioral Research, US Department of Health, Education and Welfare: 1975, *Research on the Fetus*, Government Printing Office, Washington, DC.
20. National Commission for the Protection of Human Subjects of Biomedical and Behavioral Research, US Department of Health, Education and Welfare: 1976, *Research Involving Prisoners*, Government Printing Office, Washington, DC.
21. National Commission for the Protection of Human Subjects of Biomedical and Behavioral Research, US Department of Health, Education and Welfare: 1977, *Disclosure of Research Information Under the Freedom of Information Act*, Government Printing Office, Washington, DC.
22. National Commission for the Protection of Human Subjects of Biomedical and Behavioral Research, US Department of Health, Education and Welfare: 1977, *Psychosurgery*, Government Printing Office, Washington, DC.
23. National Commission for the Protection of Human Subjects of Biomedical and Behavioral Research, US Department of Health, Education and Welfare: 1977c, *Research Involving Children*, Government Printing Office, Washington, DC.
24. National Commission for the Protection of Human Subjects of Biomedical and Behavioral Research, US Department of Health, Education, and Welfare: 1978, *The Belmont Report: Ethical Principles and Guidelines for the Protection of Human Subjects of Research*, Government Printing Office. Reprinted in the *Federal Register*, 19 April 1979, Washington, DC.
25. National Commission for the Protection of Human Subjects of Biomedical and Behavioral Research, US Department of Health, Education and Welfare: 1978b, *Ethical Guidelines for the Delivery of Health Services by DHEW*, Government Printing Office, Washington, DC.
26. National Commission for the Protection of Human Subjects of Biomedical and Behavioral Research, US Department of Health, Education and Welfare: 1978, *Institutional Review Boards*, Government Printing Office, Washington, DC.
27. National Commission for the Protection of Human Subjects of Biomedical and Behavioral Research, US Department of Health, Education and Welfare: 1978d, *Research Involving Those Institutionalized as Mentally Infirm*, Government Printing Office, Washington, DC.
28. National Commission for the Protection of Human Subjects of Biomedical and Behavioral Research, US Department of Health, Education and Welfare: 1978e, *Special Study: Implications of Advances in Biomedical and Behavioral Research*, Government Printing Office, Washington, DC.
29. National Research Council: 1988, *Mapping and Sequencing the Human Genome*, National Academy Press, Washington, DC.
30. Nozick, R.: 1974, *Anarchy, State, and Utopia*, Blackwell, Oxford, UK.
31. Parfit, D.: 1984, *Reasons and Persons*, Oxford University Press, New York.
32. President's Commission for the Study of Ethical Problems in Medicine and Biomedical and Behavioral Research: 1981, *Defining Death*, Government Printing Office, Washington, DC.
33. President's Commission for the Study of Ethical Problems in Medicine and Biomedical and Behavioral Research: 1981, *Protecting Human Subjects*, Government Printing Office, Washington, DC.
34. President's Commission for the Study of Ethical Problems in Medicine and Biomedical and Behavioral Research: 1982, *Compensating Research Injury*, Government Printing Office, Washington, DC.

35. President's Commission for the Study of Ethical Problems in Medicine and Biomedical and Behavioral Research: 1982, *Making Health Care Decisions* (with Vols. 2 and 3 appendices), Government Printing Office, Washington, DC.
36. President's Commission for the Study of Ethical Problems in Medicine and Biomedical and Behavioral Research: 1982, *Splicing Life*, Government Printing Office, Washington, DC.
37. President's Commission for the Study of Ethical Problems in Medicine and Biomedical and Behavioral Research: 1982, *Whistleblowing in Biomedical Research*, Government Printing Office, Washington, DC.
38. President's Commission for the Study of Ethical Problems in Medicine and Biomedical and Behavioral Research: 1983, *Deciding to Forego Life-Sustaining Treatment*, Government Printing Office, Washington, DC.
39. President's Commission for the Study of Ethical Problems in Medicine and Biomedical and Behavioral Research: 1983, *Implementing Human Research Regulations*, Government Printing Office, Washington, DC.
40. President's Commission for the Study of Ethical Problems in Medicine and Biomedical and Behavioral Research: 1983, *Screening and Counseling for Genetic Conditions*, Government Printing Office, Washington, DC.
41. President's Commission for the Study of Ethical Problems in Medicine and Biomedical and Behavioral Research: 1983, *Securing Access to Health Care*, Government Printing Office, Washington, DC.
42. President's Commission for the Study of Ethical Problems in Medicine and Biomedical and Behavioral Research: 1983, *Summing Up*, Government Printing Office, Washington, DC.
43. Ramsey, P.: 1975, *The Ethics of Fetal Research*, Yale University Press, New Haven, CT.
44. Rawls, J.: 1971, *A Theory of Justice*, Harvard University Press, Cambridge, MA.
45. Rothman, D. J.: 1991, *Strangers at the Bedside*, Basic Books, New York.
46. Schmeck, H. M.: 1988, 'DNA Pioneer to Tackle Biggest Gene Project Ever', New York: C1, C16, *New York Times*, 4 October.
47. US Congress.: 1983, *The Role of Genetic Testing in the Prevention of Occupational Disease* (OTA-BA-194), Office of Technology Assessment, April.
48. US Congress: 1984, *Human Gene Therapy – A Background Paper* (OTA-BP-BA-32, Government Printing Office), Office of Technology Assessment, December.
49. US Congress: 1984, *Impacts of Neuroscience* (OTA-BP-BA-24, Government Printing Office, Washington, DC), Office of Technology Assessment, March.
50. US Congress: 1985, *Reproductive Health Hazards in the Workplace* (OTA-BA-266, Government Printing Office, Washington, DC; reprinted by J. B. Lippincott, Philadelphia, PA), Office of Technology Assessment, December.
51. US Congress: 1986, *Alternatives to Animal Use in Research, Testing, and Education* (OTA-BA-273, Government Printing Office, Washington, DC), Office of Technology Assessment, February.
52. US Congress: 1987, *Life-Sustaining Technologies and the Elderly* (OTA-BA-306, Government Printing Office, Washington, DC, reprinted by J. B. Lippincott, Philadelphia, PA), Office of Technology Assessment, July.
53. US Congress: 1987, *Losing a Million Minds: Confronting the Tragedy of Alzheimer's Disease and Other Dementias* (OTA-BA-323, Government Printing Office, Washington, DC; also available through J. P. Lippincott, Philadelphia, PA as *Confronting Alzheimer's Disease and Other Dementias* and Human Sciences Press, New York, NY, under its original title), Office of Technology Assessment, April.

54. US Congress: 1987, *New Developments in Biotechnology, 1: Ownership of Human Tissues and Cells – Special Report* (OTA-BA-337, Government Printing Office, Washington, DC; also reprinted by J. B. Lippincott, Philadelphia, PA), Office of Technology Assessment, March.
55. US Congress: 1987, *New Developments in Biotechnology, 2: Public Perceptions of Biotechnology – Background Paper* (OTA-BP-BA-350, Government Printing Office, Washington, DC), Office of Technology Assessment, May.
56. US Congress: 1988, *Infertility: Medical and Social Choices* (OTA-BA-358, Government Printing Office, Washington, DC), Office of Technology Assessment, May.
57. US Congress: 1988, *Mapping Our Genes – Genome Projects: How Big? How Fast?* (OTA-BA-373, Government Printing Office; reprinted by Johns Hopkins University Press), Office of Technology Assessment, April.
58. US Congress: 1988, *New Developments in Biotechnology, 4. US Investment in Biotechnology* (OTA-BA-360, Government Printing Office, Washington, DC), Office of Technology Assessment, July.
59. US Congress: 1989, *New Developments in Biotechnology. 5. Patenting Life* (OTA-BA-370, Government Printing Office, Washington, DC; reprinted by Marcel Dekker, New York, NY), Office of Technology Assessment, April.
60. US Congress: 1990, *Neural Grafting: Repairing the Brain and Spinal Cord* (OTA-BA-462, Government Printing Office, Washington, DC), Office of Technology Assessment, October.
61. US Senate: 1990, *Departments of Labor, Health and Human Services, and Education, and Related Agencies Appropriation Bill, 1991* (101-591), Committee on Appropriations, 12 July.
62. Watson, J. D.: 1988, *NIH Press Conference: Appointment of James D. Watson to Head NIH Office of Human Genome Research, 26 September*, Videotape, Bethesda, MD: National Center for Human Genome Research; also available at the National Reference Center for Bioethics Literature, Georgetown University.
63. Watson, J. D.: 1990, 'The Human Genome Project: Past, Present, and Future', *Science* **248** (6 April), 44-49.
64. Weisbard, A. J.: 1985, *The Role of Philosophers in the Public Policy Process: A View from the President's Commission*, American Philosophical Association, Western Division, 83rd Annual Meeting, Chicago, IL. 26 April, revised 9 July.
65. Willis, D. P. (ed.): 1986, *Medical Decision Making for the Demented and Dying*, Milbank Quarterly 64, Supplement 2, Cambridge University Press, New York.
66. Working Group on Ethical, Legal, and Social Issues in Human Genome Research: 1990, *Workshop on the Introduction of New Genetic Tests*, National Institutes of Health and Department of Energy, 10 September.
67. US Congress: 1993, *Biomedical Ethics in U.S. Public Policy* (OTA-BP-BBS-105, Government Printing Office, Washington, DC), Office of Technology Assessment, June.
68. Andrews, L.B., Fullarton, J. E., Holtzman, N. A., and Motulsky, A. G., Eds, 1994, *Assessing Genetic Risks: Implications for Health and Social Policy,* Committee on Assessing Genetic Risks, Institute of Medicine, National Academy Press, Washington, DC.
69. Bulger, R. E., Bobby, E. M., and Fineberg, H. V., Eds., 1995, *Society's Choices: Social and Ethical Decision Making in Biomedicine*, Committee on the Social and Ethical Impacts of Developments in Biomedicine, Institute of Medicine, National Academy Press, Washington, DC.

PART III

CONSENSUS FORMATION IN PRACTICE

JAMES F. DRANE

DECISIONMAKING AND CONSENSUS FORMATION IN CLINICAL ETHICS

I. INTRODUCTION: THE DIFFICULTY OF ETHICAL DECISIONMAKING

Clinical ethics is about decision-making and consensus formation, both of which are sometimes difficult. Existentialists reminded us of difficulties associated with the act of deciding. Choices that involve one's own life or which affect the lives of others naturally cause anxiety. Physicians and medical ethicists are aware of this difficulty. Life and death decisions create worries and guilt and unsettling realizations of the radical limits of human life. No one makes tough medical decisions with equanimity.

Decisions in a clinical setting can also be difficult because decision-makers may feel inclined to do what is morally wrong. Temptation is not something that occurs only outside the hospital. The clinical setting not only serves up its share of unusual temptations, but sometimes intensifies and extends the ordinary ones. Even more than in ordinary life, the clinical environment serves up options and alternatives which make choosing the right thing difficult. Many clinical cases are tragic in the sense that the alternatives all seem wrong, or at least undesirable, and yet something has to be done. In many such cases the stakes are high and the consequences are both hard to determine and difficult to accept. Finally, decision-making in a context where legitimate participants are likely to have different interests, often requires mediationg conflicts and consensus building. Consensus building and decision-making are related but not identical objectives. Careful procedures promote both. Bioethics in a clinical context mediates conflict, draws participants together and is publically defensable. Another difficulty peculiar to the clinical setting derives from the fact that cases and patients are always different. In each new clinical situation, a careful analysis of relevant data is essential. The right choice in clinical cases cannot be determined in advance.

H.A.M.J. ten Have and H.-M. Sass (eds.), Consensus Formation in Healthcare Ethics, 143–157

II. CAN CLINICAL ETHICS BE ANYTHING BUT RELATIVISTIC?

Modern medicine is wedded to powerful new technologies, which creates unprecedented new possibilities. Consequently, new moral problems abound. These developments occurred just as secularization had undermined an older moral order based upon commonly held religious beliefs. Not only did each new medical advance create new moral options, but choices from among these must now be made in a climate of pluralism.

This situation leads some to despair about public ethics [3]. Any agreement about right and wrong in today's moral climate, they claim, is impossible. Radical subjectivism and relativism are taken to be inevitable. But this view seems unnecessarily pessimistic. Even when opposing views are based on belief systems which seem incompatible, negotiation and compromise are possible. Different belief systems can espouse identical ethical principles. People of good will can come to agreement about what is right, even though they disagree about ultimate meanings or the philosophical foundations of ethics. John Stuart Mill's utilitarianism is a long way from the ethics of Jesus, and yet Mill concluded that his ethics was basically the same as Jesus' Golden Rule. "In the Golden Rule of Jesus of Nazareth we read the complete spirit of the ethics of utility. To do as one would be done to, and to love one's neighbor as oneself, constitute the ideal perfection of utilitarian morality" ([8], pp. 24-25).[1] In fact the tendency of different theoretical foundations to produce a similar list of ethical standards (fidelity, autonomy, beneficence, justice, equality, respect for persons, reasonableness), undermines an undue pessimism about a possibility of overcoming radical relativism.

Persons of goodwill, including committed medical professionals, can come to agreement in most clinical situations. Given an all important commitment to doing what is right and a fairly wide agreement about guiding ethical principles, the critical issue comes down to *competent moral thinking*: moving through disciplined intellectual steps before arriving at a decision. The uncertainties associated with modern medical practice can be "tamed – if not broken," to use an equine metaphor.

III. METHODOLOGY AND DISCERNMENT IN ETHICS

If love without strategy is little more than a fleeting feeling, the same is true of ethics. The passage from moral feelings to ethics is by way of a strategy for making moral evaluations. Committed professionals will not reach the same conclusion in every clinical case but they will avoid the worst moral mistakes and come to agreed upon respectful decisions more often than not, through a process for conducting their moral evaluations. Even with a broad general agreement about moral principles, a process or methodology is required for applying the principles. It is difficult to apply the principles to concrete cases, let alone to know which one applies when two or more are in conflict. Clinical ethics is many things, but necessarily and essentially it has to be a strategy or a methodology.

Not unlike science, medical ethics must weigh, assess, analyze and study the relationships of empirical data. Unlike many schools of philosophical ethics, applied philosophy, in the form of medical ethics, is grounded in the concrete life situations where people do their living and their dying. Consequently, the practicing clinical ethicist, like the scientist, must first be a fact gatherer, and then proceed systematically to the analytic task. The competent clinical ethicist is aware of background assumptions and presuppositions at work even in this initial fact gathering stage. Objectivity is a goal in medical ethics, but it is an informed rather than a simple minded objectivity; one which takes account of the subjective dimensions even in observation and description.

No strategy or methodology can compensate for retarded ethical development or character flaws on the part of decision makers. Impulse-ridden persons, or anti-social or narcissistic personalities cannot distance themselves sufficiently from their own interest to gather objective data for evaluations, let alone to initiate actions for the benefit of patients. The decision maker in a clinical setting must at least have reached that stage of character development where response to principles and ideals is possible. Physicians are expected to operate at an "imprincipled" level of development, but instances of persons who have attained a high professional status without correspondingly high levels of ethical development are legion.[2]

A much more common obstacle to ethical discernment, however, derives from habits of coming to moral decisions without the advantages of adequate methodology. When this happens, clarity about moral judgment is missing rather than characterological capacity for discernment. Some professionals who rightfully think of themselves as decent and upstanding persons actually make decisions of great ethical seriousness in a willy nilly fashion. Others, without a systematic strategy, decide in more pragmatic ways. Some rely on authority for their moral orientation while others are confident that they themselves have an intuitive grasp of what is right. Often decisions are made according to group expectations. No adequate medical ethics can be based on these unreflective foundations. Truly professional medical ethics requires an ethical method which generates both moral discernment and consistently right judgments.

Methodology provides a framework for ethical decision making which insures that relevant data are considered. It clarifies rights and responsibilities, and reassures an ever more suspicious public that decisions important to patients and family are made with proper deliberation. But methodology does not provide infallibility. The right decision will not always be made, but the worst mistakes can consistently be avoided, and this is an important accomplishment. The authority of a methodology derives from the reasoned and respectful determinations derived from its use. Sometimes legal advice will be required before coming to an ethical decision, but most often law is satisfied when persons rightfully involved in a clinical decision are careful and systematic about how they decide. This a sound methodology can guarantee.

IV. HISTORICAL METHODOLOGIES AND EMPHASIS ON THE SITUATION

Clinical ethical strategies or methodologies are not entirely new. They are preceded by historical versions of how to arrive at defensible choices which one finds in religious ethics. Catholic moral theology was especially interested in such strategies for guiding the decisions of spiritual directors or for use in a confessional context. These historical methodologies can, in fact, be shown to have had considerable influence on the most widely used current strategies for clinical ethics.

Every method or strategy has two phases: one which directs attention to the gathering of facts; another which applies evaluative standards. A separation between the two phases usually is explicitly reflected in the model itself. The classical methodology of St. Thomas Aquinas, for example, applied standard Christian guidelines, but only after extensive attention to factual elements. He went so far as to say that human actions were right or wrong, depending on factual or circumstantial considerations.[3]

Circumstances, or the factual dimensions of a case, are considered by St. Thomas not to be accidental or of secondary importance. An ethical judgment derives neither exclusively from the structure of an act, nor from the intention which informs it. The factual dimensions, or the peculiar and particular circumstances in which a human action is performed have everything to do with its being right or wrong. Facts and circumstances are as important as evaluative standards in determining the right thing to do.

What is true of classical moral theology is also true of modern medical ethics. Medical ethics emerges from clinical contexts, and every decision is linked to a particular set of circumstances called a "case". Some forms of ethics bask in generalities and abstractions, but this is true neither of classical theological ethics nor of contemporary clinical ethics. Every helpful modern methodology gives clear prominence to the explication of medical, human, and economic factors which shift and criss cross in every clinical case.

Clinical ethics, then, is unavoidably situational, and a workable methodology must guide the explication of case particulars. Being situational, however, is not the same as being "situation ethics." Neither classical Catholic theology nor modern medical ethics are situation ethics in the sense of being radically relativistic. Objective standards and main line moral rules exist in both traditions. But in both, an act which in one circumstance would be considered killing and wrong, in another situation may neither be considered killing nor judged to be wrong (e.g., use of lethal force in self-defense).

The evaluative elements which interact with explications of facts or clinical circumstances are all attended to in classical theology: codes, statutes, precedents, ethical principles, group or individual experiences, rational arguments, cultural norms, authorities. For religious believers there are religious authorities; for secular believers, philosophical authorities. One of the most important functions of a methodology is to keep the evaluative standards connected to the factual elements. Good methodologies keep medical ethics at a distance from the danger of false generalization and keep it rooted in real life situations.

V. CASUISTRY AND CLINICAL ETHICS

A marvellous example of how methodology organizes intelligence to arrive at defensible ethical decisions is casuistry: a methodology which had its origins in Stoicism and Cicero and flourished in the 15th and 16th century, mainly among Jesuit theologians. Casuistry is defined as "the interpretation of moral issues, using procedures of reasoning based on paradigms and analogies, leading to the formulation of expert opinion about the existence and stringency of certain particular obligations, framed in terms of rules or maxims that are general but not universal or invariable, since they hold good with certainty only in the typical conditions of the agent and circumstances of action" ([6], p. 257). Theoretical assumptions (e.g., Natural Law theory) certainly were operative in casuistic thinking, as they are in modern medical ethics, (e.g., deontology and utilitarianism) but the closer one gets to clinical problem solving the farther one gets from explicit and overt theoretical considerations.

A clinical case (from *cadere* – 'to fall', therefore, 'to befall', to happen') is a statement about actions or affairs which includes reference to what classical theology called circumstances: who? what? when? where? why? how? and by what means? In the casuistic method the circumstances (*circum* around, *stare* to stand) literally stood around the core elements which were called maxims (rules or moral directives guiding moral decision making). Maxims in the sense of moral rules of thumb, much more than theory, continue to be the major evaluative elements in clinical ethics: for example, "competent patients have a right to decide"; "doctors should strive for the patients medical good"; "doctors should not take the life of a patient." Usefulness for making quick defensible decisions constitutes a maxim's "cash value." Most often more than one maxim is embedded in a case, and the clinical ethicist's role is to determine which rule really rules.

In classical casuistry certain cases served as paradigms, illustrating which maxims prevail in a given set of circumstances. As a case under consideration was shown to be similar to, or different from the paradigm case, a particular decision or rule about right and wrong was thought to be more or less certain. Everything depended upon the interplay of circumstances and maxims. Any change in circumstances will make other maxims emerge; so careful and continuing attention must always be paid to the particulars of the case. Casuistry in the past and clinical ethics now both center around cases. And circumstances are critical in each for deciding right and wrong. And in each, maxims in the sense of concrete moral directives play a central function. Decisions about right and wrong action in a clinical case are based on circumstances and justified by a maxim or a rule (like always get informed consent before initiating an intrusive diagnostic test). Casuistry, then, which Voltaire and others thought they had killed with cynical criticism, seems alive and well in contemporary medical ethics.

Clinical medical ethics shows little interest in the abstract considerations of theoretical ethics, because abstract theory has little payoff in clinical decision making. Casuistry and clinical ethics both focus on the circumstances of a case and prefer concrete directives (maxims). In the contemporary clinical context, like the historical situation in which casuistry developed, there is time pressure to decide as well as a need to justify the decision. The casuist was faced with a confessional case, or a dilemma in spiritual direction; the medical ethicist is faced with options in a medical case which must quickly be assessed and decided upon. Certain topics or considerations must always be covered, and these one finds identified in the different methodologies designed to guide clinical-ethical decisions.

VI. CONTEMPORARY AMERICAN METHODOLOGIES FOR DOING CLINICAL ETHICS[4]

Shortly after contemporary medical ethics emerged into public awareness in the late 60's and 70's, David Thomasma developed a clinical ethics program at the University of Tennessee in Memphis, U.S.A. Thomasma's program was immersed in the clinical setting and the methodology he developed for ethical decision making paralled the methodology used by doctors to make medical decisions. Thomasma distilled the moral reasoning process about cases into six steps which young clinicians were trained to follow. These steps have been slightly altered by him over the years, but essentially they are as follows:[5]

(1) Describe the medical facts of the case;
(2) Describe the values (goals, interests) of all persons involved in the case: physicians, patients, house staff, the hospital society;
(3) Determine the principal value clash;
(4) Determine possible courses of action which could protect as many of the values in the case as possible;
(5) Choose a course of action;
(6) Defend this course of action.

Thomasma defended his methodology and the need for clinicians to learn ethical reasoning procedures in a book he wrote in 1981 with Edmund Pellegrino [9].

In 1982, Albert Jonsen, Mark Siegler and William Winslade published a small volume on medical ethics written specifically for the facilitation of clinical decision making [5].[6] They reduced Thomasma's six steps to four into which they packed many complex considerations. Recognizing that doctors are used to making medical decisions which follow a certain methodology but are uncomfortable with ethical decisions, the authors discussed the reasons for physician discomforts and then moved to overcome them through a systematic approach to ethical problems. Their method was designed to provide a checklist for physicians and to guarantee that all relevant considerations are taken into account: what facts are most relevant in the case; how the facts should be organized to develop critical considerations; and how the various ethical considerations should be weighed. The four stages are as follows:

(1) Medical indications: the physician's domain;
Diagnosis, prognosis, therapeutic alternatives, clinical strategy based on risks and benefits of various courses of management, and patient particulars.

(2) Patient preferences: patient decision making based on medical indicators;
How to handle a conflict between one and two: competency considerations; overriding a patient refusal; what to do when a patient is incompetent and dying. The emphasis is on the moral priority of patient preferences.

(3) Quality of life consideration: when patients cannot decide for themselves;
When a patient is unable to make his or her decisions a surrogate must decide whether treatment creates more benefit or more burden (is the surgery, radiation, medical regime, etc. worth it?) A value is placed on features of human experience; consciousness, relationship, pain, function. Quality of life evaluations occur only when patients are unable to make judgments, their preferences are unknown, and the medical goals are limited, e.g. terminal illness, permanently unconscious patients, handicapped neonates, no codes.

(4) Socio-Economic factors: when decisions impact others;
Clinical decisions have impact beyond the doctor patient surrogate triad; upon family, limited resources, limited finances, teaching needs in medicine, the safety and well being of society. These factors are weighed last and are not given great importance in routine decisions. Increasingly, however, costs count in ethical evaluation.

In this methodology the decision maker is not only guided in his or her considerations of basic topics, but is advised when to introduce each consideration and how much weight each should be assigned. The four general topics are relatively simple but under each category many different elements and levels of ethical reflection are included.

An ethical workup designed by James F. Drane attempts to separate out the different elements and levels of discourse, as well as to show how decision makers logically proceed from one to the other [2]. Like the former model it has four main parts:

I) *Expository Phase:* Guiding the identification of relevant factual material.
1. Medical factors: diagnosis, prognosis, therapeutic options, realistic medical goals, treatment effectiveness, uncertainties associated with scientific understanding in medical practice.
2. Ethical Factors: Who is the patient and what does he or she want? What are the interests, wishes, feelings, intuitions, and preferences of patient, physicians, staff, hospital administrators, society?
3. Socio-economic factors: costs born by patient, family, hospital, HMO, insurance company, national government or local community.

II) *Rational Phase:* Guide to reasoning about the relevant data.
1. Medical ethical categories: Terms like informed consent, refusal of treatment, confidentiality, experimentation, and euthanasia create a general taxonomy for organizing data and refer to available literature. The language of medical ethics provides the tools for thinking about cases.
2. Principles and maxims: Beneficence, autonomy, respect, truth, fidelity, sanctity of life, justice, are widely accepted guides for reflection. More concrete guides come in the form of specific rules: do not prolong death, always relieve suffering, respect competent patient wishes.
3. Legal decisions and professional codes:
Paradigm legal cases guide reflection about other cases: for example, a Quinlan type case. Professional codes, updated by proclamations of professional organizations, also guide reflection.

III) *Volitional Phase*: Moving from facts and reflection to decision making.
1. Ordering the goods:
When more than one good value or interest is realizable they must

be listed according to a scale of priorities. For example, competent patient preferences have priority over physician or family preferences; in an epidemic, societal goods take preference over individual goods.

2. Ordering principles:
When principles come into conflict they are ordered according to personal belief and professional commitments. For physicians, beneficence (caring for a patient, curing, saving life, relieving pain) takes priority. Other principles are respected, but never preferred to beneficence.

3. Making decision:
Professional persons decide, with as much prudence and sensitivity as personality development permits. Special care is required whenever a decision will result in a patient's death.

IV) *Public Phase*: Preparing for public scrutiny and defense of decisions:

1. Making assumptions explicit, becoming aware of subjective factors and underlying beliefs.

2. Correlating reasons and feelings. Striving for consistency in using principles, maxims and rules.

3. Organizing arguments for public discourse. In a pluralistic society an acceptable ethic is supported by convincing reasons.

The methodologies of Thomasma, Siegler, and Drane all touch the same basic points. They differ in the explicitness with which key elements are reflected in the outline. None would disagree about any element included in the other's model. Each model attempts to provide a procedural system which can be used by clinical decision makers no matter what their theoretical beliefs are (utilitarian or deontologist, religious or secular).[7] The methodologies differ in choice of terms, ordering of topics, prominence of themes, and temporal sequences.

VII. A EUROPEAN AND LATIN AMERICAN METHODOLOGY

Hans Martin Sass, the Director of the Bochum Center for Medical Ethics in Germany, is the author of the Bochum Protocol. Jose A. Mainetti, Director of the Institute of Medical Humanities at the University of La Plata, Argentina, endorses this methodology as an alternative to "made in the USA" approaches to bioethics which according to Mainetti reflect North American culture, society, and medicine. North American medical ethics he considers to be reflective of life in the United States, which is technologized, secular, and pluralistic.

Recently, however, Mainetti sees North Americans moving toward European and Latin American styles of medical ethics. Medical traditions in Europe and Latin America are more humanistic, and their medical ethics is not so tied to deontological and utilitarian theories. Because Sass is less formalistic, theory-driven, and rule-dominated, he can (Mainetti believes) help renew medical practice.

European medical ethics is more sensitive to virtue considerations and less dominated by principles. Therefore, it needs its own methodology, which avoids even the appearance of an engineering strategy applied by technical experts to bring about socially acceptable solutions. Mainetti finds such a method in Hans Martin-Sass's Bochum Protocol which, like others we have seen, has four phases each divided into sub sections. This methodology essentially is made up of questions.

I. *Identification of scientific/medical findings:* What treatment would be best in light of the scientific medical facts?

1. General reflections:
Diagnosis, prognosis, treatment alternatives, treatment benefits or results, prognosis without treatment?

2. Special reflections:
How do treatment options with their benefits and burdens apply to this particular patient?

3. Physician's task:
Are clinical conditions such as to provide adequate treatment? Is the doctor competent? Is medical knowledge clear? Is medical ignorance recognized?

II. *Identification of ethical/medical findings:* What treatment would be best in light of ethicalmedical factors?

1. Patient Health and Well Being:
What burdens (physical, spiritual) are associated with each therapeutic alternative?

2. Patient Self Determination:
What are the patients values, attitudes, level of understanding? Can the patient's participation and decision making be respected or will it be set aside in favor of a surrogate's decision?

3. Medical Responsibility:

Can conflicts between doctor, patient, staff, and family be mediated without undermining trust, confidentiality, truth? How much clarity, certainty, doubt exists about the appropriateness of ethical categories and their inter-relationship?

III. *Case Management:* What decision is best, in light of all the above considerations?

1. What are the most acceptable options given the medical ethical findings? Is further consultation or patient transfer required?

2. What are the concrete obligations of physician, patient, staff, family in light of the treatment chosen?

3. Are there arguments against the decision? Was the decision discussed with the patient? Was the patient's consent received?

IV. *Additional Questions Regarding Ethical Evaluation:*

1. In Cases Requiring Prolonged Treatment:
Routine review of medical treatment and ethical appraisal.
Is the treatment plan flexible? Are palliative measures considered when prognosis is dismal? Is consideration of the patient's expressed and presumed wishes assured?

2. When Social Factors are Present:
Family, emotional, professional, economic complications.
Can the complications be born by the patient, family, community. How are these social factors to be evaluated vis-a-vis the medical scientific and medical ethical considerations?

3. In Therapeutic and Nontherapeutic Experimentation:
How does the experiment affect the medicalethical considerations? If disclosure to the patient is incomplete or is not understood, can the experiment be justified? If the patient has not consented, can it be justified? Was patient selection fair? Can the patient withdraw at any time?

VIII. SIMILARITIES AND DIFFERENCES

Similarities between the European–Latin American and the United States methodologies are many but the former shows several distinguishing characteristics. Although the principles turn out to be the same, less prominence is given to autonomy in the Bochum system. The section on patient self-determination for example is written from a physician's perspective. The Protocol asks what the doctor knows about the patient's value system, attitudes, understanding. This approach leads to the question: "To what extent can the patient be taken into consideration or to what extent can he (she) be completely set aside"? [10] Such a phrasing would not be imaginable in an American methodology, where the cultural emphasis on patient autonomy is strong. In the Bochum Protocol, treatment decisions are primarily the doctor's responsibility and the physician is asked to consider discussing issues with patients and then to decide whether or not to follow patient preferences. The same difference in the place given to patient autonomy vis-a-vis the physician's values is seen in the section on experimentation. The Protocol asks questions about justifying research, when the patient is not informed or has not given consent to participation. In the USA, any medical research without very informed consent would be ethically and legally indefensible.

The Bochum Protocol is as formalistic and technical as the North American models, but displays elements that are not found in the latter. More explicit references are found to epistemological issues, e.g., "What important factors is the doctor ignorant of?" and "Are the medical concepts sufficiently clear?" Taking medical ignorance and unclarity into consideration is not quite North American style. Young doctors, in fact, work hard during medical training to develop an impression of certainty and self confidence (some would say infallibility). Any suggestion that patients should be apprised of cognitive or technical limitations would be thought of as unrealistic and even anti-therapeutic by most North American doctors.

The four methodologies move from the more simple to the most complex. Thomasma's model provides the simplest outline of critical issues. Siegler lists only four topics but includes under each many more complex considerations. Drane unpacks some of the complex issues and organizes them according to an epistemologically progressive schema. The Bochum Protocol incorporates clinical, ethical and epistemological issues and, in addition, covers different clinical settings. Evaluation of the different models requires a testing of effectiveness and usefulness to practitioners. How well are the crucial elements kept before the decision maker? Is the decision maker sensitized to critical problem areas? Are potentially unrecognized issues signalled? Is the model pragmatic or clinically workable? Personally, I see advantages and disadvantages to each model and suspect that practitioners will decide for themselves which one works best and how the most workable model can be improved upon.

IX. CONCLUSION

Different approaches to ethical decision making are more than understandable. European and Latin American medicine is more humanistic in the sense that medical training continues to include philosophy of medicine, history of medicine, medical anthropology, and now medical ethics. It is easier with such a background to consider a less technical and more philosophically sophisticated decision making approach. North American doctors and medicine, on the other hand, are more clinically focused and less appreciative of philosophical issues in the practice of medicine. Philosophy and the history of medicine have practically vanished from the curriculum in the United States. Its educational strengths lie more along pragmatic lines. Ideally, a methodology would be both clinically practical and philosophically sophisticated. As cooperation between North American, European, and Latin American ethicists moves forward, both goals will come closer to realization.

Department of Philosophy
Edinboro University of Pennsylvania
Edinboro, U.S.A.

NOTES

1 The very same point was made by Thomas Hobbes (see [4], pp. 144-145, who is even further removed from religious foundations of ethics.

2 See Kohlberg's levels of ethical development [7].

3 "Actiones humanae secundum circumstantias sunt bonae vel malae" [12].

4 The methodologies in the article are shortened versions or outlines of the originals.

5 The workup cited is an update of the 1978 version which Thomasma uses at Loyola University Stritch School of Medicine (see [13]).

6 Subsequently Mark Siegler wrote a separate article on the methodology (see [11]).

7 Howard Brody provides an outline of different methodologies, depending on one's theoretical belief: utilitarian, deontologist, etc. See [1].

BIBLIOGRAPHY

1. Brody, H.: 1981, *Ethical Decisions in Medicine*, 2nd edition, Little, Brown & Co., Boston.
2. Drane, J. F.: 1988, 'Ethical Workup Guides Clinical Decision Making', *Health Progress* (December).

3. Engelhardt, H. T.: 1996, *The Foundations of Bioethics*, 2nd ed., Oxford University Press, Oxford.
4. Hobbes, T.: 1839, *Leviathan*, in W. Malesworth (ed.), *The English Works of Thomas Hobbes*, Volumes II and III, John Bohn, London.
5. Jonsen, A., Siegler, M. and Winslade, W.: 1982, *Clinical Ethics*, McMillan Publishing Company, New York.
6. Jonsen, A.R. and Toulmin, S.E.: 1988, *The Abuse of Casuistry*, University of California Press, Berkeley.
7. Kohlberg, L., Turiel, E. and Lesser G. (eds.): 1971, *Psychology and the Educational Process*, Scott, Foresman, Chicago.
8. Mill, J. S.: 1987, *Utilitarianism*, Longmans, Green, London.
9. Pellegrino, E. and Thomasma, D.: 1981, *A Philosophical Basis of Medical Ethics*, Oxford University Press, New York.
10. Sass, H.M. and Viefhues, H.: 1989, 'Bochumer Arbeitsbogen zur medizinethischen Praxis', in H.M. Sass (ed.), *Medizin und Ethik*, Reclam, Stuttgart, pp. 371-375.
11. Siegler, M.: 1982, 'Decision-Making Strategy for Clinical-Ethical Problems in Medicine', *Archives of Internal Medicine* 142 (November).
12. Thomas Aquinas: *Summa Theologica I II*, Question 18, Article 3.
13. Thomasma, D.: 1978, 'Training in Medical Ethics: An Ethical Workup', *Forum of Medicine*, December.

HANS-MARTIN SASS

ACTION DRIVEN CONSENSUS FORMATION

I. OVERVIEW

We seek consensus for different reasons. One reason is theory-driven, the other action-driven. The theory-driven demand for and the formation of consensus is frequently initiated by a need to orient oneself within a particular worldview or a unified scientific or metaphysical interpretation of one's social and natural environment. The action-driven demand for consensus is often provoked by social needs for mutual aid, cooperation, and various other forms of action within the social and natural environment. In complex situations characterized by appeal to technical, economic, or social facts or requirements (e.g., in biomedical research or patient care), consensus-formation procedures are primarily action-driven, since they cannot be based on technical criteria alone; rather, they must be based on a prudent mix of normative considerations and expertise. In simple cases such as those common to laboratory medicine, there will only be a modest or minimal need to appeal to moral maxims or other non-technical criteria. Cases that involve medical interventions at the beginning and end of human life (e.g., decisions for or against intensive care for the dying, or decisions for or against abortion), will require appeal to non-technical criteria, like moral maxims.

A worldview- or theory-driven demand for consensus in intellectual matters is evident in religious, ideological, and philosophical bases of argumentation. Achieving total or acceptable levels of consensus among a large number of people who support a particular theory usually involve certain basic values shared by those holding these views, thereby supporting their system of reference and encouraging interaction among those sharing this perspective. Theory formation that is rooted in an individual's need or desire to resolve theoretical conflicts or problems is often driven by internal factors. But when worldviews or theoretical concepts drift too far from social or political "reality," then external factors can more easily become instrumental in shaping the requirements of theory-generated consensus formation.

An action-driven demand for consensus does not usually require theoretical or ideational coherence; no unifying theory is needed to

H.A.M.J. ten Have and H.-M. Sass (eds.), Consensus Formation in Healthcare Ethics, 159–173

develop a particular tool, to use it correctly, or to produce and "market" a certain good. Certain conditions, however, are needed to achieve consensus: (1) the identification of mutual or reciprocal interests and (2) the clarification of benefits, costs, and risks of certain actions. For action-driven consensus one need only "acceptable" levels of theory that are based on specific and adequately allocated moral maxims rather than abstract theories. We need only to possess acceptable (comfortable?) levels of technique and technical knowledge, specific and adequately allocated expertise rather than complete professional knowledge of (say) engineering, biomedicine, or economics. Therefore, in action-driven consensus formation, it is important to determine what types of theory and knowledge are needed to achieve a consensus, as well as what knowledge should be included. In theoretical matters, people disagree on the importance of various principles, either because agreement is seen as of no value for the particular action, or because agreement cannot be reached for reasons of technical uncertainty (or theoretical controversy), or because of the prevalence of high technical, moral, cultural, or even political risk ([12,14]).

Action-driven consensus formation activity is essentially non-theoretical. It is not "theory free," but can be called "theory modest" ([2], p. 39-44). On the other hand, theory-driven consensus may be defined as non-practical, and in some cases even "hostile" to the realities of the world, which it intends to surpass, exceed, or transcend. Theories sharpen those skills necessary to analyze and shape realities; empirical and practical activities provide information to be used in theory formation. Members of groups that espouse certain ideologies and religious views, that seek political or "indoctrinating" power are often driven to action and apply their concepts as action tools. In contrast, those who rely on scientific approaches often retreat into ideological theory, being interested more in conceptual orientation than in practical action – even though the ideology itself might be used for action-oriented consensus formation.

Special consensus-formation procedures, required and shaped by what Kant called the "primacy of practice" [Praxisprimat], become complicated in the world of theories or values as well as in the worlds of engineering, economics, and biology, because they must integrate both theoretical and practical reasoning.

From the Kantian perspective one would not abandon normative reasoning in favor of situational utility. But normative reasoning, just like reasoning in engineering, has to be differentiated and fine-tuned in order to be practically effective where technical, economic, ethical, legal, and social factors shape a given situation. Whitbeck describes in detail how practical ethics can learn from engineering design [20]. Whatever system of reference is used – and in a pluralistic society quite a number of different systems of belief are used by persons who have to cooperate in a given situation – it is important that individual and collective values, principles, visions, and expectations be differentiated and balanced according to the complexity of the situation. (I have elsewhere described such a normative, but differentiating approach under the rubric "differential ethics" [12,13,14].) In the words of Jennings, a consensus-oriented ethics model is "a practice of discourse rather than a body of objective knowledge" ([8], p. 447). Again, as far as the practice of rational and intuitive reasoning in discourse is based in a normative conception of democratic moral agency, consensus-oriented ethics is not simply utilitarian; rather, it justifies itself on normative grounds as long as the consensus is shared by and based in "a fabric of related categories – community, citizenship, membership, consent, openness, accountability, and communication" ([8], p. 462).

In what follows, I shall discuss two different models of conflict resolution: (1) scenarios which might allow for content-related societal consensus, and (2) scenarios in which societal "dissensus" rather than consensus exists and where, nevertheless, law and order and the recognition of individual conscience and self-determination have to be reconciled. In particular, I will assess (1) the function of consensus as one among other instruments for conflict resolution and interaction, (2) the use of ethical and technical rules and requirements, principles and facts, in moral scenario assessment, and (3) the relevance of the "subsidiarity principle" for societal consent in accepting dissent (i.e., appealing to self-determination and individual conscience when individuals confront unresolved societal controversies).

II. CONFLICT RESOLUTION: COMPROMISE, CONSENSUS, AND COOPERATION IN TRUST

It is necessary to differentiate three levels of praxis-oriented interaction that have a tendency to cause conflict: the expert-expert interaction, the lay-expert interaction, and societal interaction. Conflicts can be of a purely technical or ethical nature, or may be rooted in a power struggle.

When one discusses conflicts between the ideologies of the disciplines of ethics, science, and medicine, one is primarily concerned with the identification, analysis, evaluation, and resolution of ethical conflicts that arise in patient care and in biomedical research. Here one can draw on experience with risk assessment procedures in other types of technical and moral risk because the methods of risk analysis, containment, and resolution are of a similar nature in different areas of personal and professional action.

Cases of conflicting expert opinion in transdisciplinary cooperation can occur because of differences in the experts' knowledge, risk analysis outcomes, situational assessment, and professional goals and targets. In expert-lay interactions, conflicts may be rooted in deficient information about facts and/or values, deficient technical and non-technical risk assessment, inadequate consent or contractual arrangements, diversity of worldviews and goals in life, and judgments of quality of life. Experts may in many cases be considered non-experts or lay persons in each others' field of expertise. Trust that is based on a foundation of experience, information, or cooperation is a prerequisite for expert-lay cooperation, and the resolution of conflict among experts who assess various situations. Professionalism-in-trust may lead to communication-in-trust and thus to cooperation-in-trust among professionals.

It is questionable, however, whether expert committees have the moral authority paternalistically to assume that their specific judgments or actions should have validity for the treatment of individual patients or for social or cultural assessment. As Veatch has pointed out, technical experts are not *per se* experts in public policy or in making substitute moral assessment for other individuals [19]. But overlapping expert competence is usually a good means for defining situations more precisely and to assess technical as well as moral risk germane to different courses of action. Therefore, expert consensus as well as expert "dissensus" on risk factors and risk-rewards can balance different policy options and are valuable instruments "to help decision makers come to their own conclusions about what they ought to do" ([18], p. 409); but, in the words of Tong, "second-hand ethics is no substitute for first-hand ethics in a community of equally autonomous moral agents" ([18], p. 415). Indeed, public policy and regulatory policy of European Union bureaucratic bodies, repeatedly come under scrutiny and are heavily criticized, as those who formulate them seem to not understand the role of moral communities and individual decision makers to make moral choices based on their first-hand understanding of risks and responsibilities.

In expert-lay relations communication and cooperation usually begin with beneficence-in-trust and compliance-in-trust. Non-expert goal description, maxim definition, or authority-based personal leadership will facilitate the setting of an agenda for cooperation within which professional expertise usually plays a role. However, there are certain rights and responsibilities that come with having professional expertise as well as civil liberties that may not be compromised, some of which are norms or moral principles. (I only note here that "scenario evaluation," case analysis, and questionnaires [4,5,10,12,15,17] are often useful instruments for guiding professionals, lay persons, or groups when they find themselves in the role of decision makers).

The development of political or social conflict resolution techniques can be observed in the history of political theory and practice and of social ethics and social research. Traditionally, people have often wondered: on which maxims of public policy can we, as educated citizens, agree? Participants involved in consensus formation have also introduced a second question: which principles can be ignored in the consensus-formation agenda without shredding the basic fabric of trust and cooperation in a society that is highly diversified and constituted of a multiplicity of cultures? While a familiar and proven tool in political conflict resolution, the liberal strategy of reducing conflict by increasing individual awareness of risk and professional responsibility has not been effective in contemporary discussions. Clearly, the moral and cultural limits of the application and development of biomedicine and other forms of technology require further exploration ([12,14]).

In the professional as well as the political setting, requirements for consensus may vary from consensus on data acquisition, data interpretation, procedural rules, moral, technical, or regulatory criteria and lists of priority goals, to consensus on leadership and decision-making processes. Cooperation can be based on either one of these forms of consensus and may include either a consensus not to agree on certain goals of the interaction or to agree on a compromise on specific procedures or the instruments to be used. Ideational consensus or consensus on private or professional virtues may be helpful, but cannot be made a prerequisite for cooperation in multicultural societies among consenting adults in their roles as citizens, experts, consumers, clients, and patients. What are the minimal requirements in multicultural societies and pluralistic democracies for a consensus that will allow for trust-based cooperation?

Compromise may be a good instrument with which to resolve conflicts between technical and economic principles such as efficiency, durability, and price. Compromise may also be a means of balancing moral principles such as autonomy with professional paternalism or the freedom of the individual in tension with the interest of the common good. But how far can contractual compromises go that do not incorporate a recognition of the need to protect basic moral and human rights without violating human dignity, the common good, or the prevailing environmental ethos?

Finally, there are scenarios in which for generations (individuals, groups, churches, lawyers, legislators, and regulators) have fought, persecuted, and even killed people in the name of forging consent, rather than allowing for, even tolerating, dissent. These are the scenarios where conscience clauses in national laws or the individualization rather than generalization of moral judgement and moral responsibility will be the best instrument for conflict resolution, also the one of last resort ([13,14]).

Whatever the means are of achieving cooperation in complex personal, professional, social, and political methods of facilitating cooperation – compromise, consensus, contract, or individual free choice – none of these instruments will be efficient if trust-based interaction cannot be achieved. So, the preeminent requirement for all instruments to be used in professional interaction, whether consensus-based or not, is trust [14]. Identification of moral problems and the achievement of consensus will not be fully successful if highest priority is not given to the establishment of trust between and among experts, experts and lay persons, and citizens in their private and professional roles.

III. MORAL SCENARIO ASSESSMENT

Beginners in assessing moral scenarios often confuse ethics with emotion, or with the direct application of cognitive sentences to practical matters. Both forms of moral malpractice are responsible for the fact that outright cruelty or moral mismanagement have occurred in the name of "ethics." "Emotions are necessary but not sufficient resources for ethical decision-making," according to Fletcher and other experienced decision makers in biomedical ethics. Bioethical reasoning may start and return frequently to the emotional level but is meant to achieve higher levels of judgment than "gut-level" responses ([5], p. 11). General ethical principles – such as justice, liberty, and security are not sharp and specific enough for the assessment of specific moral scenarios that require the application of very distinct technical and ethical criteria. Micro-allocation and mix-allocation of moral principles in single case situations as well as in scenario analysis will, according to R.M. Hare, have to "include an account of the critical as well as of the intuitive level. The critical level is that at which we select the principles to be used at the intuitive level, and adjudicate between them in cases where they conflict" ([7], p. 36).

In the assessment of moral scenarios one has to work with "mid-level principles" – such as the various notions of *nil nocere, bonum facere*, and the best interest of the patient ([9], p. 75; [14]). Following Aiken and Fletcher ([1], [5], p. 11-13), we can describe four steps in evaluating complex moral scenarios: (1) the emotional level at which feelings of sympathy, solidarity, or hate are expressed, (2) the moral level where the central question is – "What ought to be done?" (3) the ethical level, where one asks – "Why should it be done?" and (4) the post-ethical level that includes an assessment of the degree to which the actions are ethical. The professional ethos thus has to transgress the realm of sympathy and emotional attachment, preferences and dislikes, and has to deal with differentiations in case analysis and scenario assessment, including various additional organizational, economic, and political issues. All this can only be achieved by the highest degrees of rational, clear, and distinct analysis, evaluation, and judgment. Technical risk expertise without ethics is blind; ethics without technical differentiation is empty. Both have to be integrated as closely as possible in professional training and in professional service. Along those lines, the authority of decisions made by healthcare ethics committees, teams of experts in engineering, or other consensus-formation bodies cannot simply be based on a majority vote, as arguments have to be weighted, not counted.

Mid-level principles are, for some philosophers, important in ethics, just as they are in technical or economic consensus formation processes. General ethical principles such as autonomy, beneficence, and non-maleficence have to be related to specific cases and scenarios. General ethical principles can be understood as value commodities, which will become useful only in "semi-finished" versions. Semi-finished moral principles are mid-level principles such as codes of professional conduct, free speech (a semi finished form of autonomy). Amalgamated value products include informed consent (professional responsibility and client autonomy), the patient's best interest (*nil nocere plus bonum facere*), most guidelines and regulations, and legal, organizational, technical and political instruments and procedures. In scenario analysis one deals with the scenario-adequate mix and balance of semi-finished value components, which one may call maxims. In individual cases, adequate micro- and mix-application of mid-level principles can facilitate the articulation of certain values; they form the fabric of case assessment and patient care ethics in medicine and of good client-provider relations in professional services. We may disagree about the "commodity" level, but such a disagreement might be totally irrelevant for consensus formation or at least trust-based cooperation formation at the "end product" level.

Systems of law, codes of professional conduct, regulations, contracts, and market rules are typical scenarios for mix-allocation of different principles. Special mix-value scenarios can be described for the physician-patient paradigm (patient autonomy, physician responsibility), the producer-consumer paradigm (consumer need, demand, stimulated demand, producer interest, cash flow, market share, product safety, efficiency, stability of the product), issues of justice, concern for the environment, and regulation, the government-citizen paradigm (civil protection and social-redistribution conflicts regarding civil rights, market forces weakened by regulation, personal risk competence weakened by paternalism).

Moral reasoning in scenario development and case analysis follows four steps: analysis, assessment, review, and management. More precisely, we can differentiate 10 steps: (1) detailed analysis of single risk components, (2) assessment and grouping of single risk components, (3) development of risk reduction and risk avoidance strategies, (4) definition and discussion of options for action, (5) selection of one option, (6) assessment of arguments against the selected option, (7) modification and confirmation of the decision, (8) managing moral risk, (9) periodic re-assessment of the scenario or case, (10) *ad hoc* re-assessment of the scenario or case.

When moral or technical consensus is the preferred goal of interaction, it is prudent to discuss simple cases first, complicated or extreme cases later. In bioethics it would be moral malpractice to start with ethically extreme or worst-case scenarios. In human fertility treatment, for example, in-vitro fertilization (IVF) is first discussed by referring to cases of infertile couples, then to cases of single mothers or sperm-, egg-, and embryo-donation, and only then to extreme cases such as zygote splicing or special demands for lesbian insemination by same-sex donor which might have (in the worst case), moral costs that far exceed benefits. By discussing relatively uncomplicated cases one first tries to keep ethics simple, even at the risk of higher technical and economic costs.

Action Guides for bioethical decision making and consensus formation have been developed and can be used as instruments in establishing *prima facie* maxims and principles for a rapid orientation to specific scenarios. In order to work effectively, they will have to be implemented by specific, differentiated scenario-evaluating checklists or other additional instruments.

Action Guides for patient care and good physician-patient relations must be centered around the mid-level principles, like *primum non nocere*, beneficence, patient autonomy, and professional responsibility. While professionals and lay persons may agree to such an approach in cases of urgent need for lifesaving interventions, the "consensus-based" cooperation between patient and physician is more desirable in all other cases. In bioethics one seems to find the same mid-level principles whether they are used to develop virtue-based action guides, such as the "Posthippocratic Oath" formulated by Pellegrino and Thomasma ([11], p. 205ff) or the intertwined virtue tables I employed elsewhere [14], a principle-based ethics as developed by Beauchamp and Childress [3], covenants or contracts between physician and patient as proposed by Veatch ([19], p. 327-330), or questionnaires ethics modules, or an ethics meta-language for interactive expert systems.

An Action Guide for the moral management of clinical research would have to incorporate mid-level moral criteria specific to human experimentation.

These action guides should not be conflated with the "worst case scenario" method, where the course of action and consensus depend on the worst possible outcome. Action guides use the "differentiated scenario' in developing the highest standards for reducing technical and moral risk, and in fostering cooperation and consensus. In contrast, worst-case scenario consensus procedures represent evaluative and moral malpractice, as they do not allow for detailed and differentiated evaluation; they represent "defensive ethics" at the cost of patient care and the further development of medical and moral resources [12]. Scenario evaluation is exploratory and anticipatory, essential not only for consensus development in anticipatory technology assessment prior to the introduction or development of new technology or procedures, but also in setting moral strategies for specific patient care parameters in hospitals or wards. Scenario assessment can also be used in training students, staff, and professionals in the formation of procedures for fostering consensus.

Procedures for forming consensus using methods for assessing moral scenarios allow for the micro-allocation and mix-allocation of technical as well as moral facts and principles. Existing classification guides that elucidate various technical scenarios in patient care should be implemented only when accompanied by moral parameters. The ICD classification scheme of the World Health Organization, for example, would make a good candidate for the introduction of ethical criteria into the context of technical criteria. Biomedical experts could design an ethics metalanguage for individual patient care that would allow the physician to introduce not only technical but also nontechnical issues into her consensus-oriented interaction with the patient. The integration of moral maxims into technical descriptions would provide for a more patient-oriented prognosis and therapy. The achievement of this integration on a global scale would require intensive cross-cultural studies of attitudes towards health risks, pain, disease, and quality-of-life issues. Interactive electronic expert systems could be made morally compatible with ethical experts or, at least, could become ethically more acceptable, if either modules of moral software, maxims, and principles, or a moral meta-language were integrated into the expert system. The software of integrating differentiated particles of specific technical and moral essentials in the assessment of moral scenarios follows established and proven patterns of technology assessment and risk analysis in engineering and economics.

IV. CONFLICT RESOLUTION: CONSENSUS AS TRUST IN DISSENT AND THE PRINCIPLE OF SUBSIDIARITY

Consensus in basic values and convictions is, as we discussed, not necessary for conflict solution as long as (1) all parties agree on an acceptable course of action, (2) the course of action is based on semi-finished mid-level principles, and (3) there is a fair chance one will find answers to questionnaires which can be agreed upon by all or most parties. Mid-level principles can be accepted as a common basis for consensus formation even if they are supported by different modes of reasoning or different basic religious, metaphysical, or utilitarian models – such as Jesus showed in the parable of the Good Samaritan (Luke 10) where he underlines the idea that the principle of "neighborly love" or (in modern parlance) "solidarity" can be supported by quite a variety of different world-views and denominational convictions.

Here I shall only briefly discuss both the moral issue of abortion and the conceptual challenges surrounding criteria for declaring (diagnosing) death in the present-day clinical setting. The Harvard formulation of 1968, defining the death of an individual (in additiona to the traditional heart-beat/ circulation/respiration/pulsation criteria) as "the cessation of all functions of the entire brain, including the brain stem" [13]. This found overwhelming support in most nations of Judeo-Christian-European origin reflecting both the Greek humanist tradition as well as the Bible, wherein one accepts the distinction between mortal body and immortal soul; that the soul derived from and would return to the Creator. So it was not difficult to conclude, as did Pius XII and many others, that the immortal soul might already have left the mortal, dying body in cases of prolonged and irreversible coma [13]. As "brain-based criteria" for declaring death had found wide support for over 25 years, it was not thought inconsistent to assume that a similar formula might be adopted to establish a consensus concerning the moral recognition and legal protection of unborn human life, also based principally on criteria of (total) brain functioning. In particular, the biblical narrative of delayed, divine animation by the Creator following the gestation of Adam's body (Gen. 2,7), combined with the Aristotelian-Thomist teaching of animation, suggested that brain-based criteria in the context of human death might very well achieve public support as did brain-death criteria, when first proposed, pertinent to the human fetus. In 1989, I suggested that the translation of biomedical facts of neuro-maturation, in particular the formation of post-mitotic cortical cells forming synapses after the 70th day post conception, into bioethical assessment of the dignity of unborn human life could result in a formula which would legally protect and morally recognize unborn human life from the moment of earliest synapsis formation in the future cortical plate, prior to which there would be no "brain tissue" which could be protected, only isolated and permanently subdividing and migrating neurons [13]. But this proposal, even though at least as strongly embodied in dormant values and concepts as the prevailing brain-death proposal did not find the same support. The pro-life and pro-choice camps still fight their battles with little hope of reconciliation (or even winning the war), reflecting no apparent interest in consensus.

During the last decade (at least in Germany), the previously unquestioned acceptance of brain-based criteria for diagnosing death has also eroded, and strong forces – in particular from some Protestant schools of thought – have vehemently called for alternatives to the uniform brain-death definition. As in the case of abortion, ideological differences and the calls of individual consciences and personal responsibilities based on religious traditions, cultural attitudes or *prima facie* moral instinct inhibit general and uniform solutions and create new targets. These failed attempts to base societal cooperation and the recognition of the individual's conscience on commonly shared premisses and suggest that we need to seek alternatives to material consent and replace it with procedural consent and the willingness to respect each others' individual responsibility, based on deeply rooted and dearly held visions and conviction, but different ones. The fact that one side of the aisle are persons as responsible and as considerate as on the other side, marks the specific postmodern scenario of multiculturalism and the different priorities set by individuals and groups regarding visions and values, which remain widely shared but differently ranked and balanced.

It was the Thomist principle of subsidiarity which in similar situations of threats of social uniformity and totalitarianism in social ethics (in particular Fascist and Stalinist social theory and social policy in the 1930s), that was re-appreciated and further developed in order to allow for a rich variety of approaches in caring for the poor, sick, and disadvantaged, as the 1931 encyclical 'Quadrogesimo Anno' (article 79) stated: whatever the individual can do, should not be done by the state or by social institutions or uniform services; what small communities can do, should not be taken away from them by larger bodies [14]. If we would carry the subsidiarity principle from the fields of social ethics, where it had worked very well, into the fields of medical ethics, it might help to reduce unwarranted and perhaps unnecessary societal conflict over general principles and general strategies, and promote individual responsibility and the individual conscience as the prime source for making crucial choices where society in general lacks the power and authority to impose uniform solutions ([13,14]).

If we translate the reasoning of the subsidiarity argument from social ethics into bioethics, we can reshape the abortion debate and formulate an action-driven conscience model which would hold that whenever theologians, philosophers, legislators and regulators, lawyers and bureaucrats disagree, and whenever society does not have and does not find strong and convincing commonly shared values, then those "closest" to the choices and decisions (in the case of the abortion debate, pregnant women and their families) should be the prime moral agents for making crucial choices ([13,14]). Similar flexibility should be given to the individual determination and definition of personal criteria for death, and for withholding life-sustaining treatment, if concerned individuals wish to do so [14].

A strongly shared consensus to live with "dissensus" on the most important issues of personal life, vision, belief, and expectation, is the foundation for free and open democratic societies that respect human dignity: dignity of the individual's conscience, self-determination, and self-responsibility. To strive for one-for-all objective solutions regarding issues on which we hold different opinions, is inappropriate and violates the preconditions of discourse-based democratic societies. Therefore, beginning-of-life and end-of-life moral issues will best be treated by invoking the subsidiarity principle and by recognizing the individual person's conscience as the source of moral agency. Value-based questionnaires are helpful instruments for improving and empowering individual decision making [9] as are narrative materials [10], and the use of advance medical directives [10].

V. CONCLUDING WORD: A DECLARATION REGARDING TOLERANCE

A societal consensus to live with societal dissent regarding e.g., the morality of abortion and criteria for declaring personal death, could read as follows: "We, the people, residing in a nation rich in cultural traditions, and in support of self-determination and individual responsibility, respect the fact that responsible individuals assess moral situations and challenges differently. We therefore hold: (1) the dignity of persons is unalienable; (2) unborn human life must be legally protected and morally recognized (a) from the moment of implantation, (b) from the beginning of integrating brain circuitry (estimated to be around the 10th week of gestation), or (c) viability of the fetus; (3) human life is no longer in need of legal protection and does not require moral recognition, when, and only when, all functions of (a) the cardiovascular system, (b) the entire brain including the brain stem, or (c) the neocortex have irreversibly ceased to function; (4) the State may, in order to provide for security of its citizens, as well as for medical and moral protection and safety, determine one of the possible positions as the default position most likely agreed upon by a majority of its citizens; and the state must allow for conscience clauses that allow individual citizens freely and responsibly to choose other positions and act accordingly; and (5) the State must provide for consultation services (protecting free moral choice within the limits of the law) bearing on reproductive choice."

Such a declaration would reflect a societal consensus on the importance of tolerating and accommodating individual moral choice and conscience. In short, a specific doctrine and procedure need to be in place to permit an individual to choose, e.g., if a woman is dissatisfied with society's default position she must have recourse to act on her firmly held moral convictions.

It was Spinoza, who in his *Tractatus Theologico Politicus* of 1670 called for this approach (which we recognize as consensus formation though he stressed toleration and subsidiarity) when he proffered the thesis that toleration in ideational and religious matters (replacing century-old models of consensus formation by indoctrination, war, and terror) would neither result in anarchy nor immorality, but a stronger, peaceful society sustained by self-determined and self-responsible citizens. After introducing the principle of subsidiarity in bioethical consensus-formation procedures, it could one day happen that in some scenarios (among them probably the abortion issue and the issue of recognizing patients' advance directives), accepting dissent will not be the second best solution in conflict resolution but the preferred or even the best one.

Kennedy Institute of Ethics
Georgetown University
Washington, D.C., U.S.A.

Zentrum für Medizinischer Ethik
Ruhr Universität
Bochum, Germany

BIBLIOGRAPHY

1. Aiken, H.D.: 1982, *Reason and Conduct*, Knopf, New York, NY.
2. Arras, J.D.: 1991, 'Getting down to cases: The revival of casuistry,' *The Journal of Medicine and Philosophy* **16**, 29-51.
3. Beauchamp, T.L. and Childress, J.F.: 1983, *Principles of Bioethics*, 2nd ed. Oxford University Press, New York, NY.
4. Drane, J.F.: 'Methods of clinical ethics,' in this volume, pp. 143-157.
5. Fletcher, J.C.: 1990, *Basic Clinical Ethics and Health Care Law*, Ibis Publishers, Charlottesville, VA.
6. Gibson, J.M.: 1990, 'Reflecting on values,' *Ohio State Law Journal* **51**, 440-452.

7. Hare, R.M.: 1981, 'The philosophical basis of psychiatric ethics' in S. Bloch, P. Chodoff (eds.) *Psychiatric Ethics*, Oxford University Press, Oxford, UK.
8. Jennings, B.: 1991, Possibilities of consensus: Toward democratic moral discourse, *The Journal of Medicine and Philosophy* **16** (4), 447-463.
9. Jonsen, A.R. and Toulmin, S.: 1988, *The Abuse of Casuistry*, University of California Press, Berkeley, CA.
10. Kielstein R., Sass H.-M.: 1993, Using stories to assess values and establish medical directives, *Kennedy Institute of Ethics Journal* **3**, 303-325.
11. Pellegrino, E.D. and Thomasma, D.C.: 1988, *For the Patient's Good: The Restoration of Beneficence in Health Care*, Oxford University Press, New York, NY.
12. Sass, H.-M.: 1987, 'Philosophical and moral aspects of risk and manipulation,' *Swiss Bioetech* **5** (2a), 50-56.
13. Sass, H.-M.: 1994, *Die Würde des Gewissens und die Diskussion um Schwangerschaftsabbruch und Hirntodkriterien*, Zentrum für Medizinische Ethik, Bochum, Germany.
14. Sass, H.-M.: 1996, The clinic as testing ground for moral theory: A European view, *Kennedy Institute of Ethics Journal* **6** (4), 351-355.
15. Sass, H.-M. and Viefhues, H.: 1989, 'Bochumer Arbeitsbogen zur medizinethischen Praxis,' in H.-M. Sass (ed.) *Medizin und Ethik*, Reclam, Stuttgart, pp. 371-375.
16. Schoelmerich, P. and Thews, G.: 1990, *'Lebensqualität' als Bewertungskriterium in der Medizin*, G. Fischer Verlag, Stuttgart, Germany.
17. Spicker, S.F.: 1990, *Medical-Ethical Questionnaire in Diagnosis*, Zentrum für Medizinische Ethik, Bochum, Germany.
18. Tong, R.: 1991, The epistemology and ethics of consensus, *The Journal of Medicine and Philosophy* **16** (4), 409-426.
19. Veatch R.M.: 1991, Consensus of expertise: The role of consensus of experts in formulating public policy and estimating facts, *The Journal of Medicine and Philosophy* **16** (4), 427-446.
20. Whitbeck, C.: 1996, Ethics as design: Doing justice to moral problems, *Hastings Center Report* **26** (3), 9-16.

K.W.M. FULFORD

DISSENT AND DISSENSUS: THE LIMITS OF CONSENSUS FORMATION IN PSYCHIATRY

Psychiatry is a contentious and problematic discipline: diagnostic difficulties, some empirical, others conceptual; ethical dilemmas, such as those related to the psychiatrist's wide-ranging powers of compulsory treatment; medicolegal problems, extending to uncertainty even about the remit of the psychiatrist as an expert witness; and a potential for abuse, for political as well as personal ends – all these amount to barriers to consensus formation which are out of all proportion to those arising in the rest of medicine. Small wonder, then, that there should be disagreement about the very concept of mental illness, about whether conditions conventionally regarded as mental illnesses are properly so-called [6].

The purpose of this chapter is to argue that diversity of view in psychiatry is to a degree not only irreducible but legitimate.[1] As we will see, there has been some progress in recent years towards reducing the degree of disagreement in psychiatry over questions of diagnosis. Taken together with the success generally of scientific medicine, this has encouraged the view that in the end all disagreements in psychiatry, all differences of opinion, will be resolved by appropriate scientific advances. It will be argued here that this is a false extrapolation. There is further progress to be made, certainly. But underpinning the diversity of view in psychiatry is the diversity of human nature itself, a diversity which, far from eliminating, we should seek to accommodate.

I. REDUCIBLE DIAGNOSTIC DISAGREEMENT

The early development of psychiatry in the nineteenth century was characterised by a wide variety of approaches to the diagnosis of mental disorders. As Kendell has put it, it seemed that every self-respecting alienist, and certainly every professor, had his own classification [23]. With the publication of Kraepelin's *Lehrbuch*, however, a degree of order began to emerge. Focussing attention on the symptoms and course of mental disorders, Kraepelin, together with Jaspers [21] and others in the

H.A.M.J. ten Have and H.-M. Sass (eds.), Consensus Formation in Healthcare Ethics, 175–192

German phenomenological tradition, developed a classification which remains the basis of those used today. Even so, wider acceptance of a single classification had to wait until after the end of the second world war when the newly-formed World Health Organization published the International Classification of Diseases. Early versions of the psychiatric chapter of this classification were much criticised, but the eighth edition [35] was adopted by most member states. The main rival to ICD at this stage was the American Diagnostic and Statistical Manual [2]. By a process of convergent evolution, however, the ICD and DSM, though still differing in detail, have now become broadly similar [36].

The situation today is quite different from that in the nineteenth century, therefore. Important areas of disagreement and difficulty remain, in primary care [22] for example, and in liaison psychiatry [25]. Moreover, psychiatrists continue to be highly idiosyncratic in their diagnostic formulations: the details of the psychiatric classification still vary from textbook to textbook; and the diagnostic categories actually employed in practice are often different again from any of these. But we have now a widely agreed "basic" classification [13], developed from a well-established phenomenology of mental illness, and adopted at least in mainstream hospital psychiatry in many countries throughout the world.

This shift to consensus in psychiatry has been driven by the adoption of a scientific approach, more or less consciously emulating that of physical medicine. Thus the relative nosological disorder of the nineteenth century was no more than a reflection of early attempts to develop aetiological "disease models" of mental illness, inspired by the germ theory and other scientific medical successes. Kraepelin's contribution was in effect to recognise that all such theories were premature. With his emphasis on the symptoms of mental disorder he took psychiatry back to a descriptive stage of disease classification, equivalent to that adopted by Sydenham and others for physical medicine in the seventeenth century. The ICD and DSM, too, have been developed on the same theory-free principle. The innovative ICD-8 was based on a report by Stengel [29] in which he emphasised the importance of clear descriptive definitions of terms. This in turn was inspired by a paper by the philosopher of science, Karl Hempel, distinguishing the descriptive and explanatory stages in the development of a science [17]. Indeed much of the phenomenological tradition in psychiatry, from Jaspers through to the standardised mental state examinations of present-day psychiatric research, has been conceived essentially as a basis for the development of a scientific psychology [21]. It is the momentum of this approach which has led psychiatrists to identify so strongly with the scientific model of physical medicine. It is true that progress towards an explanatory science of psychiatry has been disappointingly slow. Certain physical treatments have been found to be valuable for limited (though serious) kinds of disorder; and recent developments in brain-imaging and in molecular genetic techniques look promising. But we have as yet established few mental disease entities equivalent in heuristic power to, say, pneumococcal pneumonia or B12-deficiency anaemia. Nonetheless the progress towards consensus on a descriptively-based psychiatric classification, together with the success of a broadly scientific approach, has sustained the belief that psychiatry can develop simply by following in the tracks of physical medicine. Past successes point to future successes, so the hope goes.

There are however two reasons for doubting that this hope will be fully realised. The first is that even as a descriptive science the progress of psychiatry towards consensus is less dramatic viewed from outside than from within. Those most closely associated with the development of our present classifications are sometimes inclined to believe that everything is more or less settled. But we have already noted that in areas such as primary care and liaison psychiatry – in which after all a great deal of psychiatric pathology is to be found – these classifications are inadequate. Moreover, even in hospital psychiatry they are sustained in part by training and by peer reinforcement. The PSE, for example, one of the standardised mental state examinations, was actually designed to model the practice of leading psychiatrists [34]: and the DSM categories are more or less mandatory in the USA, both for election into the profession by examination and for obtaining payment for services from insurance companies. Philosophers of science have of course long pointed out that pressures of this kind are important in the establishment of paradigms in all sciences [18]. But in psychiatry the paradigms are far from settled. Medical psychiatry can distance itself from certain of its competitors, from psychoanalysis, say, or from sociology, by claiming that these are not "scientific". But there are competing scientific models as well. Psychologists, in particular, though no less "scientific" than doctors, adopt a quite different approach to diagnosis; and psychologists, indeed, by the ultimate medical test of practical efficiency, have been highly successful in developing powerful treatment methods [15].

So even as a science psychiatry can be far from confident that it is on the right track in following physical medicine. But the second reason for doubting this is stronger still. What it amounts to is that there are non-scientific – specifically, evaluative – elements in the conceptual structure of medicine which, though largely unproblematic in physical medicine, are highly problematic in psychiatry. The very progress of physical medicine as a purely scientific discipline has thus been dependent on a simplified view of the concepts of illness and disease. It is with this simplification, and the corresponding complications in psychiatry, that we will be concerned in the next section.

II. IRREDUCIBLE DIAGNOSTIC DISAGREEMENTS

Consistently with the scientific self-image of medicine, the conventional line among doctors has been that the concepts of illness and disease are, at heart, essentially factual in nature. It is acknowledged that there are ethical aspects to medicine. But the doctor's expertise as a doctor, it is thought, is, or should be, restricted to purely scientific disease concepts defined, ultimately, by reference to norms of bodily and mental functioning. This "science-based" view of medicine is not confined to doctors, of course – a particularly clear version of it has been developed by the philosopher Boorse [4], for example – but it is the view they conventionally hold. A quite different line has often been taken by those outside medicine, however. The sociologist Peter Sedgwick, for instance, argued that, appearances notwithstanding, even in physical medicine disease concepts are essentially value-laden [28]. This conclusion has often been taken by doctors to be hostile to medicine. The very possibility of medical science is assumed to be dependent on a science-based, rather than value-based, view of the medical concepts [4]. A value-based view of medicine, however, can be understood as reflecting no more than a different perspective. The science-based view reflects the doctor's perspective. From this point of view what matters is scientific knowledge of disease: it is with this that doctors are mainly concerned, not only in clinical work but in medical education and research. From the patient's point of view, on the other hand, what matters, ultimately, is the experience of illness, an experience which is nothing if not (negatively) value-laden. Understood in this way, therefore, the science-based and value-based views of medicine, far from being hostile to each other, should be fully reconcilable, and recent work in the philosophy of medicine has been directed towards just this end ([8,27]). There are many complications here, of course, as to the particular conceptual elements which are emphasised, and the precise ways in which they are taken to be related. But a useful basic approach involves thinking of the attribution of disease as involving two steps: step 1 is the plain description of someone's condition, in itself a value-free process; step 2, however, is the value-laden process of judging that condition to be good or bad, healthy or diseased. To the extent that science itself is value-free, this keeps intact the scope of medical science, concerned as it is with the descriptively-defined conditions of step-1. But with the evaluations of step-2, it remains a value-based view. The net result is a more rounded or complete view of the conceptual structure of medicine. To the facts emphasised in the doctor's science-based view are added the value judgements inherent in the patient's experience of illness. And by extension, to the analysis of disease concepts in terms of failure of functioning, is added an analysis of the experience of illness in terms of failure of action, or incapacity. It is this more complete picture of the conceptual structure of medicine which is the key to understanding the limits of consensus formation in psychiatry.

Fact and Value

The idea that there is an evaluative, as well as factual, element in the conceptual structure of medicine, is perhaps less surprising in psychiatry that in physical medicine. For the concept of mental illness is prima facie more value-laden than that of physical illness. This indeed has been a major focus of the debate in the literature about the validity of mental illness. Both sides in the debate have been much influenced by the science-based view of medicine. Opponents of mental illness, like Szasz, have taken its value-laden nature to be an indication that it is not "really" illness [30]. But supporters of the concept, too, have felt the need to apologize for and to explain away its evaluative connotations. They have regarded these as a temporary embarrassment, a reflection merely of the underdevelopment of mental science. Kendell, for example, has argued that, properly defined, mental illness can be purged of its evaluative connotations, leaving it, so far as medical science is concerned, a purely descriptive concept, as the concept of physical illness is supposed to be [24].

A view of medicine, on the other hand, which recognizes an evaluative as well as factual element in the conceptual structure of the subject, suggests a quite different interpretation of the more overtly evaluative connotations of mental illness. According to such a view, instead of mental illness being a poor scientific relation of physical illness, both concepts, mental and physical, properly reflect the properties of illness as a value term. Thus as Hare [14], Urmson [31] and others have pointed out, the strength of the evaluative connotations of all value terms, even those like good and bad, may vary with context. 'Good', used of apples, for example, has mainly descriptive connotations (sweet, clean-skinned, etc.), while used of, say, pictures, its connotations are more overtly evaluative. So if 'illness' is a value term, then used of physical conditions it could be like 'good' used of apples, while used of mental conditions it could be like 'good' used of pictures. The mechanism of this variation, moreover, is consistent with this suggestion, and directly relevant to our understanding of consensus formation in medicine. Thus, value judgements are made according to criteria which are descriptive in nature – the criteria for a good apple are that it should be sweet, clean-skinned, etc. These criteria, however, are not fixed. They vary from person to person, and from occasion to occasion. And differences in the strength of the evaluative connotations of a value term are a direct reflection of this variation. Most people most of the time take sweet, clean-skinned, etc., apples to be good, whereas there are no such widely agreed descriptive criteria for a good picture. Hence while "good picture" remains an overtly evaluative term, the evaluative connotations of "good apple" have become largely eclipsed by its descriptive connotations. Applying this to medicine, then, we see that similar differences are to be found in the ways in which we typically evaluate mental and physical phenomena. Pain, for example, as a typical symptom of physical illness, is more or less universally regarded as at best a necessary evil. But anxiety, say, as a typical symptom of mental illness, is evaluated differently by different people and in different circumstances. Some people actually enjoy horror films and dangerous sports, and these are not rare perversions as the enjoyment of (at least extreme) pain would be. Just as with the term 'good', then, if 'illness' is a value term, the more overtly evaluative connotations of mental illness would be no more than a reflection of the way in which we evaluate mental symptoms such as anxiety, and the more descriptive connotations of physical illness would be no more than a reflection of the way in which we evaluate physical symptoms such as pain.

It is important to see that the criteria involved here have nothing to do with illness as such. They are first and foremost criteria by which we evaluate phenomena such as anxiety and pain in general, and only thereby and secondarily criteria for the particular value judgements involved in taking these phenomena to be symptoms of illness. We will return to this point in the next section, looking at how illness is marked out from other kinds of negatively evaluated condition. It is also important to emphasise that the criteria involved here are not, as such, scientific criteria. The same criteria may indeed be involved at both stages in the attribution of disease, serving first to define a condition (step-1, descriptive) and second as the criteria by which that condition is judged to be good (step-2, evaluative). But they remain distinct. Moreover, whereas the way in which we divide up mental and physical phenomena into distinct conditions is subject to scientific "discoveries", our evaluations, including our evaluations of mental and physical phenomena, are not. Rottenness in apples, to revert to our non-medical example, could be defined by, say, biochemical criteria (step-1), and in most contexts an apple which met these criteria would be a bad apple. But this latter (step-2) judgement would be quite different in, say, the context of cider making. Here a rotten apple would be a good apple.

All this puts a quite different construction on the evaluative connotations of mental illness. As we saw earlier, in a science-based view these connotations are in one way or another prejudicial to psychiatry. But on the present account they show psychiatry to be a more complex and demanding subject than physical medicine. In terms of our two-step model of disease attribution, what this amounts to is that in physical medicine – given the relative uniformity of our step-2 evaluations of phenomena such as pain – diagnostic difficulties are largely confined to step-1 (descriptive). But in psychiatry – given the relative variability of our evaluations of phenomena such as anxiety – diagnostic difficulties may arise not only at step-1 (descriptive) but also at step-2 (evaluative).

This account also puts a quite different construction on why physical medicine has been able to develop as though it were a purely scientific subject. According to a science-based account, this is because the concept of physical illness is, indeed, a value-free scientific concept. According to the account now outlined, it is, rather, because the evaluative (step-2) element in the attribution of physical illness is (largely) uncontentious and hence can (generally) be ignored. The concept of physical illness, that is to say, can generally be assumed to be value free, because the evaluative element in its meaning is by and large unproblematic. This is the simplification, then, available to physical medicine but not to psychiatry, referred to in the introduction.

In physical medicine the difference between these two accounts is (mainly) academic. But in psychiatry this is far from the case. As we have seen, a science-based account implies that psychiatry will only make progress scientifically by eliminating step-2, by defining away the evaluative connotations of mental illness. A value-based account, on the other hand, shows this to be a potentially dangerous misconception. According to such an account, all that is required (in principle) for the progress of psychiatry as a science is clarification of the proper scope of the descriptive element, the step-1 element, in the conceptual structure of the subject. There is no requirement to eliminate the evaluative, step-2, element. Indeed to the extent that this element is genuinely problematic in its own right, denying or suppressing it is likely to compound the difficulties, practical as well as scientific, presented by the subject. I have shown elsewhere that this is no mere theoretical possibility, explaining as it does the particular vulnerability of psychiatry to abuse [9]. But before going into this in more detail, we must look first at an important extension to the value-based account itself.

Action and Function

The argument of the preceding section may be thought to underestimate the scientific approach to medicine. In concentrating, in effect, on symptomatically defined conditions it is in tune with the present descriptive stage in the development of disease classification in psychiatry. And at this stage, it may be admitted, value judgements could well come into the diagnosis of mental disorders. But descriptive classifications, this line of argument might continue, were intended, by Kraepelin no less than by the authors of modern classifications, merely as a preliminary to the development of classifications based on aetiology, on the causes of illness. And once we understand the causes of mental illness, once we have adequate ways of investigating brain functioning, once, that is, we can "test" for schizophrenia or depression in the same way that we can test for anaemia, then the evaluative element in diagnosis should disappear. This, after all, is how physical medicine, along with many another science, has developed, from a descriptive to an explanatory stage. The brain, to be sure, is a more subtle organ than any other, presenting a greater technological challenge than livers or kidneys, say. But there is surely nothing other than complexity in the way of the eventual emergence of causal- rather than descriptively-based classifications of mental disorders, and hence of diagnostic procedures in psychiatry which are value-free.

This argument thus extends the science-based view to an explanatory stage. However, an explanatory science-based view is open to objections of broadly the same two kinds as those outlined in the last section in respect of a descriptive science-based view. In the terms of our two-step model, these are step-1 and step-2 objections. Thus, in respect of step-1 the argument underestimates the difficulties inherent in the development of psychiatry as a science. Hopes for an explanatory science of psychiatry are generally combined with a model of science as instantiated by biology: brains, as it was put a moment ago, differ from kidneys and livers merely on a scale of complexity. But there are reasons for believing that in its development as a science, psychiatry may well have more in common with physics. Psychiatry shares with quantum theory, in particular, puzzles about the relationship between the observer and the observed, puzzles which physical medicine, like much of biology, has been largely able to ignore [6]. So, as with psychiatry as a descriptive science, the assumption that as an explanatory science it will follow slavishly in the path of physical medicine, is far from secure.

Even leaving aside, however, this (admittedly speculative) point, the step-2 or evaluative element in the attribution of disease turns out to be as important in relation to an explanatory science-based view as it was in relation to a descriptive, and for similar reasons. Consider physical medicine first. Here, as we have seen, it has been possible by and large to ignore the evaluative element in the diagnosis of disease essentially because, although present, it is largely unproblematic. And the basic point to be made here is that if this is true at a descriptive level it will remain true at an explanatory. Thus, if a symptomatically-defined condition (angina, say) is unproblematically pathological, then the underlying causal changes in bodily structure and functioning (in the heart) will be unproblematically pathological as well. Hence the illusion that the symptomatically-defined disease concepts of physical medicine are value-free carries through seamlessly to the underlying causally-defined concepts. In the case of causally-defined concepts, moreover, this illusion gives rise to another. For as Ayer argued, the notion of cause implies merely a tendency to produce a given effect [3]. The underlying "lesion" of ischaemic heart disease is a lesion because of its tendency to produce angina. Such lesions, therefore, once recognised, are properly called pathological even though, at the time in question, the patient has no actual symptoms. The fact that causally-defined concepts have, in this sense, a conceptual life of their own, allows powerful diagnostic short cuts – direct ECG testing for ischaemic heart disease, for example. But this has given rise to the illusion, our second illusion, that illness in general is defined by the presence of underlying pathological changes in bodily structure and functioning. Whereas, as we have seen, changes in bodily structure and functioning are on the contrary marked out as pathological by a tendency to cause illness.

It is this second illusion, then, that lies behind medical hopes for a value-free explanatory science of psychiatry. Even in physical medicine, however, its effects can be pernicious – the all-too-common response of doctors to the patient whose symptoms have defied explanation in terms of established pathology, is that there is "nothing wrong". In psychiatry, given the present descriptive stage of development of the subject, this would leave precious little in the way of genuine mental illness. But the psychiatric counterpart of the argument just given for physical medicine shows, anyway, just how illusory the illusion itself is. Thus, if a symptomatically-defined psychiatric condition (hysteria, say) is only contentiously an illness, then the (putative) underlying changes in bodily structure and functioning (in the brain) can only be contentiously pathological as well. Even granted developments in the brain sciences, therefore, there is not the same scope for diagnostic short cuts in psychiatry as there is in physical medicine. To think otherwise amounts to making "cause" identical with "cause of pathology". Yet such is the power of our second illusion that it is no less than this identification which is often made by psychiatrists. In forensic psychiatry, for example it is generally assumed, tacitly, that the mere demonstration of a correlation between some aspect of bodily structure or functioning and a given behaviour is sufficient to mark out criminal behaviour as pathological [19].

Just what would be sufficient to mark out behaviour as pathological is a large question. The point made here, however, is that this question should be addressed directly, by analysis of the concept of illness, rather than by reference to underlying changes in bodily structure and functioning. This is well illustrated by the occasional case in which criminal behaviour is due to pathology [10]. What is shown by such cases can be summed up by saying that disease may mitigate but only illness can be an excuse. Thus frontal lobe brain damage, perhaps due to a brain tumour, may result in socially disinhibited behaviour and hence in sexual or other offenses. And in these circumstances the appropriate response would not be to punish the person concerned but to arrange for them to receive treatment. This is not because of the presence of a lesion as such, however. A court might deal leniently with an offender who has a tumour, whether in the brain or elsewhere. Adverse circumstances, medical or non-medical, can be adduced in mitigation of an offence. But for a lesion to be an excuse in law, for it to absolve from liability altogether, it must have had an effect on the offender's capacity for responsible action. And this takes us firmly into the conceptual area of illness. For as a number of authors have argued, whereas many disease concepts can be analysed in terms of impairment of functioning, the experience of illness is more appropriately analysed in terms of incapacity, or impairment of intentional action ([6,27]).

The question of definition raised at the start of the last paragraph is thus large indeed. For with the analysis of illness as action-failure we are drawn into all the complications of the philosophy of action, and through these into areas of deep metaphysics such as the nature of persons, the rationality of belief, and the problem of freedom of the will. We should not resist this, however, since these complications are clearly present in medicine itself. It is persons (not bodies or parts of bodies) who fall ill; illness, as incapacity, impairs our freedom (this is why illness excuses); mental illness, in particular, is irrationality (delusion is the paradigm symptom of mental illness as an excuse in law). There is a general sense in which these complications show the scope for diversity of view in psychiatry. It is with mental illness, after all, rather than mental disease concepts, that we are mainly concerned in this area of practice. But there is also a sense in which these complications tie this diversity of view specifically to the value judgements inherent in medicine. For what is at issue in psychiatry, in the forensic cases already referred to, for example, and in questions of involuntary treatment, is the distinction between one kind of negatively evaluated condition and another – "mad versus bad", for example, in forensic psychiatry, and "mad versus sad" in the involuntary treatment of suicidal patients. All in all, then, the move from a descriptive to an explanatory science-based view, far from undermining the significance of the evaluative logical element in medicine, actually endorses it.

III. IMPLICATIONS FOR CONSENSUS

This account of the conceptual structure of medicine may seem a recipe for diagnostic chaos. A science-based view of medicine holds out the prospect (in principle at least) of agreement. But if, as has been argued here, value-judgements are inseparable from medical diagnosis, if the medical concepts are inherently value-laden, then, it might be said, what hope is there for consensus? This would be overly pessimistic, however. In the first place, as noted earlier, the descriptive element in diagnosis, the step-1 element, is left undisturbed by this account. In physical medicine, to the extent that step-2 or evaluative considerations can be ignored, this in effect leaves undisturbed a large part of our established diagnostic procedures. And even in psychiatry, the recognition of the importance of step-2 considerations does nothing to diminish the importance of step 1. It is no less important to attend meticulously to the symptoms of which the patient complains, to adopt clear descriptive definitions which allow these symptoms to be reliably identified, and to establish statistical groupings of symptoms into genuine syndromes. The very complexities of step-2 in psychiatry, let alone the relative absence of adequate knowledge of brain functioning, make all these step-1 considerations more, not less, clinically important. And indeed the PSE, and other standardised mental state examinations, have shown that the degree of consensus which it is possible to achieve in this descriptive part of psychiatric diagnosis, is entirely comparable with that achieved in physical medicine [5].

In the second place, however, even step-2 in psychiatry, involving as it does evaluations, will not be chaotic. This is simply because evaluations themselves are not chaotic. The basis of the present account is the observation that we disagree over our evaluations of phenomena such as anxiety more than over our evaluations of phenomena such as pain. But this is not to say that we disagree widely even over our evaluations of anxiety. The point can be put in the positive by saying that to the extent that we agree in our evaluations of phenomena such as anxiety, so we will agree in our evaluations of these phenomena as symptoms of (mental) illness. Urmson, indeed, taking this a stage further, has pointed out that agreement of this kind allows in some circumstances what he calls "conventional" definitions of value-terms, that is the adoption by convention of an agreed set of descriptive criteria for a value-judgement made for a particular purpose. One of his examples is the criteria defined by the Ministry of Agriculture for the purpose of grading apples [31]. A value-based view of medicine suggests that we should understand many of the categories in our standard classifications of mental disorders similarly. For these categories, too, are defined by sets of (largely) descriptive criteria, criteria which are adopted by convention for the purpose of making diagnoses. On the present account, then, these criteria are to be understood as descriptive criteria adopted by convention for the value-judgements entailed by the diagnosis of mental disorders.

The psychological stability of our evaluations has also led to a more radical philosophical position. It has been argued that it should sometimes be possible to go beyond mere conventional definitions of value terms to the actual reduction of evaluations to descriptions, the relevant descriptions thus entailing the value judgements in question [32]. I have suggested elsewhere that this "moral descriptivism" offers a more secure foundation for a science-based view of medicine than the eliminavist positions generally adopted [6]. Instead of seeking artificially to eliminate evaluation from the conceptual structure of medicine, moral descriptivism allows the continued use of illness and disease as value terms, while at the same time providing in principle for logically determinate descriptive criteria for the value judgements they express. Lowered life expectancy and reproductive capacity, for example, instead of displacing the evaluative element in the meaning of disease (as proposed by Boorse [4] and others), would entail the value judgement "disease": hence disease could still be used with evaluative connotations, but the criteria for its use, instead of being subject to individual variation, would be fixed. Moral descriptivism is thus highly attractive as a basis for consensus formation in medicine. For it preserves a value-based view of the subject while reducing the individual variation normally inherent in our value judgements to the orderliness of descriptive science.

Moral descriptivism, however, is unlikely to be helpful in psychiatry. Its persuasiveness depends on the existence of cases in which, as with apples and pain, we tend to agree in our evaluations. But psychiatry, according to the view presented here, differs from (many) other areas of medicine precisely in that we tend to disagree in our evaluations of phenomena like anxiety, and, for this reason, over the diagnosis of mental illness. Hence just where consensus in medicine is most elusive (i.e., in psychiatry), the prima facie case for a moral descriptivist approach is least persuasive. We may pursue such an approach, nonetheless. Indeed, the arguments of Boorse and others, referred to a moment ago, are open to reinterpretation in just this way [6]. In seeking to identify criteria for disease which, though perhaps neither necessary nor sufficient, are nonetheless uncontentious at least in physical medicine, these authors can be understood as attempting to establish a set of descriptive criteria which everyone must, logically must, accept. But even if this could be done, its application to psychiatry would at best be irrelevant to all those cases which are genuinely contentious in everyday practice. At worst it would amount to consensus through the imposition of medical criteria (derived from physical medicine) over those of other professionals, and indeed over those of patients themselves and their relatives. Consensus would thus be achieved, but artificially, by reinforcing the medical hegemony – a *pax iatrogenica*, to coin a hybrid phrase.

IV. THE CASE FOR DISSENSUS

Thus far the arguments of this chapter have been concerned with the negative issue of the extent to which disagreements can be eliminated from psychiatry. They also point to a number of positive lessons, however. The key point has been that the difficulties arising in step-2, the evaluative step in the diagnosis of mental illness, are no more than a reflection of the variability of our evaluations of phenomena, such as anxiety, which, when they are symptoms of illness are usually symptoms of mental illness. A first lesson then, is that we should drop the usually pejorative stance towards mental illness. Its evaluative connotations, far from making it a poor scientific relation of physical illness, can now be seen to be a faithful reflection of a logical property which, as a value term, it shares with other value terms, including "good".

The evaluative connotations of mental illness, moreover, underscore important lessons about medicine as a whole. Outside the context of hospital medicine, the importance of the patient's experience of illness, of "empowering" the patient, and indeed of the value-ladenness of medical diagnosis, are already part of the perspective of healthcare professionals – nurses, social workers, occupational therapists, midwives and general practitioners ([26,16]). Indeed, even in hospital medicine an exclusively scientific and technical focus is increasingly recognised to be inadequate as a basis for good clinical practice. Hospital doctors, precisely because they are responsible for ever more powerful and sophisticated scientific methods of treatment, should be the more concerned with what ought to be done as distinct from what can be done. The rapidly expanding field of bioethics is of course directed towards the ought questions of medicine. And its preoccupation with "high-tech" issues reflects the importance of these questions in hospital-based clinical practice. But as Alderson has emphasised [1], bioethics has felt constrained to come up with answers, to produce solutions to ethical problems capable of commanding the same degree of consensus as scientific answers to empirical problems. It has thus tended to fall back on excessively general principles and formulations which are often irrelevant to the muddle and disorder by which the dilemmas of the clinical situation are usually characterised. Hospital doctors, therefore, no less than other healthcare workers, must be concerned to develop the skills necessary for the successful application of their knowledge under conditions of uncertainty. It is to the necessity for these "practice skills" that the evaluative connotations of mental illness point [20].

There is a more fundamental lesson to be drawn however. For just as the uniformity of our evaluations in physical medicine suggested a moral descriptivist analysis of the medical concepts, so the variability of our evaluations in psychiatry suggests the opposite, a non-descriptivist analysis. Non-descriptivism is the view, going back to David Hume and beyond [14], that values can never, even in principle, be derived from descriptions alone – "no ought from an is", the slogan goes; to get values out values have first to be put in. It is fashionable among philosophers nowadays to regard the debate about the logical relationship between facts and values as somewhat passé. Pointing to the way in which description and evaluation are wound together into our everyday concepts, they have become impatient with analytical considerations in favour of what they take to be more immediately practical issues. But here, at least, in medicine, the analytical and the practical are one and the same. For if moral descriptivism is wrong in principle, the very drive to make psychiatry subject to the same degree of consensus as physical medicine is misdirected. Instead of reducing values to facts, let alone eliminating values altogether, our efforts should be directed towards making explicit and clarifying the respective contributions of both elements, fact and value, to the conceptual structure of the subject. We noted earlier the importance of this task in relation to scientific medicine. We see now that it is no less important in relation to clinical practice. And it is a task to which philosophy, analytical philosophy, is uniquely well suited [11]. Philosophical work in this area has indeed already begun to yield useful practical results, not only in ethics [9] and in classification and diagnosis [12], but also in a wide range of other areas, including descriptive psychopathology, medical jurisprudence, health education, and even in the development of standardised questionnaire schedules, traditionally the province of scientific sociology and psychology [7].

We should not expect progress here in a rush or all-of-a-piece. What is required is not general theorising but detailed research, clinical as well as philosophical. Neither, though, should we expect progress in this area to yield consensus. It is a reflection of the dominant scientific world-view that convergence is often assumed to be the only measure of success, even in ethics. Rationality, it is felt, must be modelled on the rationality of science [33]. A degree of agreement is essential, to be sure. As we have seen, there are a range of considerations, contingent as well as analytical, to show that we can expect this even in psychiatry. But we have also seen that there are limits to consensus, limits which are set by the variability of our value judgements. The contribution of philosophical research to medicine could thus be to develop the tools for a rational "dissensus" (even the word has to be invented!) to balance scientific consensus. This would be no merely academic contribution. For the variability of our value judgements is integral to our individuality as human beings. So long, therefore, as psychiatry is concerned with individuals, a degree of dissensus is not only legitimate but a direct reflection of good practice.

The Philosophy and Mental Health Programme
Department of Philosophy
University of Warwick
Coventry, United Kingdom

NOTE

[1] The author expresses his gratitude to Dr D. Foreman and Professor R.E. Kendell for their helpful comments on an early version of this article.

BIBLIOGRAPHY

1. Alderson, P.: 1990, *Choosing for Children: Parents Consent to Surgery*, Oxford University Press, Oxford, England.
2. American Psychiatric Association: 1994, *Diagnostic and Statistical Manual of Mental Disorders* (fourth edition), American Psychiatric Association, Washington.
3. Ayer, A.J.: 1936, *The Central Questions of Philosophy*, Penguin Books Limited, UK.
4. Boorse, C.: 1975, 'On the Distinction Between Disease and Illness', *Philosophy and Public Affairs* **5**, 61-84.

5. Clare, A.: 1979, The Disease Concept in Psychiatry in P. Hill, R. Murray, A. Thorley, (eds.), *Essentials of Postgraduate Psychiatry*, Academic Press, Grune and Stratton, New York.
6. Fulford, K.W.M.: 1989, *Moral Theory and Medical Practice*, Cambridge University Press, Cambridge, England (Reprinted 1995).
7. Fulford, K.W.M.: 1990, 'Philosophy and Psychiatry: Points of Contact', *Current Opinion in Psychiatry* **3**, 668-672.
8. Fulford, K.W.M.: 1991, 'The Concept of Disease', in S. Bloch, P. Chodoff, (eds.), *Psychiatric Ethics* (2nd edition), Oxford University Press, Oxford,
9. Fulford, K.W.M., Smirnoff A.Y.U. and Snow, E.: 1993, 'Concepts of Disease and the Abuse of Psychiatry in the USSR', *British Journal of Psychiatry* **162**, 801-810.
10. Fulford, K.W.M.: 1993, 'Value, Action Mental Illness and the Law', in J. Gardner, J. Horden, and S. Shute (eds.), *Criminal Law: Action, Value and Structure*, Oxford University Press, Oxford, England, pp. 279-310.
11. Fulford, K.W.M.: 1993, 'Bioethical Blind Spots: Four Flaws in the Field of View of Traditional Bioethics', *Health Care Analysis* **1**, 155-162.
12. Fulford, K.W.M.: 1994, 'Closet Logics: Hidden Conceptual Elements in the DSM and ICD Classifications of Mental Disorders', in J.Z. Sadler, M. Schwartz and O. Wiggins (eds.), *Philosophical Perspectives on Psychiatric Diagnostic Classification*, Johns Hopkins University Press, chapter 9.
13. Gelder, M.G., Gath, D. and Mayou, R.: 1989, *Oxford Textbook of Psychiatry* (second edition), Oxford University Press, Oxford, England.
14. Hare, R.M.: 1963, 'Descriptivism', *Proceedings of the British Academy* 49, 115-34. Reprinted in Hare, R.M.: 1972, *Essays on the Moral Concepts*, The Macmillan Press Ltd, London, England.
15. Hawton, K., Salkovskis, P.M., Kirk, J. and Clark, D.M.: 1989, *Cognitive Behaviour Therapy for Psychiatric Problems: A Practical Guide*, Oxford University Press, Oxford, England.
16. Helman, C.G.: 1981, 'Disease Versus Illness in General Practice', *Journal of the Royal College of General Practitioners* **230** (3), 548-52.
17. Hempel, C.G.: 1981, 'Introduction to the Problems of Taxonomy', in J. Zubin (ed.), *Field Studies in the Mental Disorders*, Grune and Stratton, New York, pp. 3-22.
18. Hesse, M.: 1980, *Revolutions and Reconstructions in the Philosophy of Science*, The Harvester Press.
19. Hill, D.: 1962, 'Character and Personality in Relation to Criminal Responsibility', *Medicine, Science and the Law* **2**, 221-232.
20. Hope, R.A. and Fulford, K.W.M.: 1993, 'Medical Education: Patients, Principles and Practice Skills', in R. Gillon (ed.), *Principles of Health Care Ethics*, John Wiley and Sons, Chichester, England, pp. 697-709.
21. Jaspers, K.: 1913, *General Psychopathology*, 1963 (transl. Hoenig, J. and Hamilton, M.W.), Manchester University Press, Manchester, England.
22. Jenkins, R., Smeeton, N., Marinker, M. and Shepherd, M.: 1985, 'A Study of the Classification of Mental Ill-Health in General Practice', *Psychological Medicine* **15** (2), 403-9.
23. Kendell, R.E.: 1975, *The Role of Diagnosis in Psychiatry*, Blackwell Scientific Publications, Oxford, England.
24. Kendell, R.E.: 1975, 'The Concept of Disease and its Implications for Psychiatry', *British Journal of Psychiatry* **127**, 305-15.
25 Mayou, R. and Hawton, K.: 1986, 'Psychiatric Disorder in the General Hospital', *British Journal of Psychiatry* **149**, 172-190.

26. McKee, C.: 1991, 'Breaking the Mould: a Humanistic Approach to Nursing Practice', in R. McMahon, and A. Pearson (eds.), *Nursing as Therapy*, Chapman and Hall, Suffolk, England.
27. Nordenfelt, L.: 1987, *On the Nature of Health: an Action-Theoretic Approach*, D. Reidel Publishing Company, Dordrecht, The Netherlands.
28. Sedgwick, P.: 1973, 'Illness – Mental and Otherwise', *The Hastings Center Studies* I (3), 19-40 (Institute of Society, Ethics and Life Sciences, Hastings-on-Hudson, New York).
29. Stengel, E.: 1959, 'Classification of Mental Disorders', *Bulletin of the World Health Organisation* **21**, 601-63.
30. Szasz, T.S.: 1960, 'The Myth of Mental Illness', *American Psychologist* **15**, 113-118.
31. Urmson, J.O.: 1950, 'On Grading', *Mind* **59**, 145-69.
32. Warnock, G.J.: 1967, *Contemporary Moral Philosophy*, The Macmillan Press, London and Basingstoke.
33. Williams, B.: 1985, *Ethics and the Limits of Philosophy*, Fontana, London, England.
34. Wing, J.K., Cooper, J.E. and Sartorius, N.: 1974, *Measurement and Classification of Psychiatric Symptoms*, Cambridge University Press, Cambridge, England.
35. World Health Organization: 1978, *Mental Disorders: Glossary and Guide to Their Classification in Accordance with the Ninth Revision of the International Classification of Disease*, WHO, Geneva.
36. World Health Organization: 1992, *The ICD-10 Classification of Mental and Behavioural Disorders: Clinical Descriptions and Diagnostic Guidelines*, WHO, Geneva.

GEBHARD ALLERT, GERLINDE SPONHOLZ,
HELMUT BAITSCH AND MONIKA KAUTENBURGER

CONSENSUS FORMATION IN GENETIC COUNSELING: A COMPLEX PROCESS

> Genetic counseling is a communication process that deals with the human problems associated with the occurrence, or the risk of occurrence, of a genetic disorder in a family. This process involves an attempt by one or more appropriately trained persons to help the individual or family to, (1) comprehend the medical facts, including the diagnosis, the probable course of the disorder, and the available management; (2) appreciate the way heredity contributes to the disorder, and the risk of recurrence in specified relatives; (3) understand the options for dealing with the risk of recurrence; (4) choose the course of action which seems appropriate to them in view of their risk and their family goals and act in accordance with that decision; and (5) make the best possible adjustment to the disorder in an affected family member and/or the risk of recurrence of that disorder [4].

There is general agreement among genetic counselors about this definition. It was formulated in 1972 during a workshop on Genetic Counseling in Washington, D.C. ([4,15, 16]).

I. THE COMPLEX FIELD OF INTERACTION IN GENETIC COUNSELING

Processes of communication are characterized by two main categories: "Process" implies a sequence of time, "Communication" means an interaction of persons. Communication (persons) and process (time) are characteristics of all genetic consultations. At least two persons (i.e., client/counselee and counselor) are involved. In general the consultation will last one or two hours. Almost every genetic counseling is very complex having also a previous history (pre-consultation process) and a long history following the counseling (post-consultation process).

It is hardly possible to determine the exact duration of the pre-consultation process; however, one can conclude from the contents of the consultation that in many cases this pre-story lasted a long time,

H.A.M.J. ten Have and H.-M. Sass (eds.), Consensus Formation in Healthcare Ethics, 193–207

sometimes several years. Generally, many persons are involved in this previous history: partners, children, grandchildren and relatives, close friends or acquaintances, institutions and authorities, health insurance and other insurance companies, etc. In this period processes of communication take place covering general or specific topics that lead to or are closely related to genetic counseling.

In the clients' post-consultation story which also might take a long time, one might find again processes of communication where several or many persons are involved: clients and their partners, children, grandchildren, relatives, close friends or acquaintances, institutions and authorities, health insurance or other insurance companies.

Even the counselor has his personal and professional story. That is to say, he has a personal biographic background with an individual system of values, his individual idiosyncrasy and abilities of communication. He underwent a specific professional training in a specific school and he gained experience. The counselor is influenced by the norms and moral concepts of his scientific community, his working atmosphere, his religious or philosophical orientation ([12,14]). Before the conversation, the counselor made himself familiar with the client's specific problem. He worked out the indication with the help of the available data. He elaborated a more or less rigid strategy concerning the form and contents of the consultation. Often, he has communicated with members of his institute, the laboratory, the clinic, the referring doctors, institutions for long-term care and handicapped people, psychologists, psychotherapists and social workers. The counselor's post-consultation period is again marked by an interactive process of communication which takes time and involves communication with different persons. The clients are informed about the findings and these findings are explained. Scenarios of decisionmaking are discussed with colleagues of the institute, of the clinic, and with the referring doctors. Even other persons concerned (children, relatives) may take part in this process. The post-consultation process is not finished at this moment; often more questions are asked by the clients or other persons and institutions involved in the counseling. Moreover, the contents of the conversation become the object of reflection or analysis in conversations with colleagues, in so-called case studies and in supervisions, where style and contents of the consultation may undergo a critical analysis.

In this complex communication process of genetic counseling, with different pre- and post- stories, with internal and external interconnections, problems of informed consent appear at various points and in different personal configurations. Figure 2 gives an overview of these different levels of the process of communication and the required consensus. In addition to the three dimensions showed in the graph, the dimension of time is also present in all consensus processes. This graph illustrates that the main level of the complex field of genetic counseling is related to other dimensions. The processes of decision-making in genetic counseling are especially influenced by the frames of personal and social-political norms and values. The different fields on the main level show the persons directly involved in genetic counseling (clients, counselor) and the other groups, institutions, authorities, etc. The arrows symbolize the interaction between these groups. The field "Human Genetics" describes the complex interactions of formal and informal aspects of the scientific community as well as research and molecular biology, institutes of human genetics and genetic counseling services. The field "Medical Practice" represents practitioners, gynecologists, pediatricians and others, clinics and social workers. The field "Patients Groups" covers the many self-support initiatives. Field "Social Context": means occupation, social position, economic factors. Field "Public Information" covers: radio, television, press, informative material and lectures. The field "Family" covers relatives and friends.

In order to illustrate the intensity and complexity of the required processes of consensus formation, four genetic counseling sessions are given as examples. They are from the Department of Clinical Genetics of the University of Ulm. The conversations were partly taped and transcribed ([3,9,13]). After a short description of the problems arising during the sessions we focus on the interactions dealing with the problems of consensus formation.

Case 1: Clinical Diagnosis: Muscular Atrophy Werdnig-Hoffmann

A couple, both 40 years old, come to genetic counseling. The basic problem is a muscle disease, that caused the death of one son of the family some years before. This genetic counseling was initiated by the husband who has informed himself about this disease. The information he got from a newspaper article made him believe that this disease was a sex-specific form of a muscular dystrophy and that his three daughters might bear the risk of being carriers. His wife does not share this view. She is convinced that the child died as a result of a vaccination. The counselor informed the client in detail that the disease in question is a recessive muscular atrophy Werdnig-Hoffmann and that the risk of recurrence of this disease in his daughters' offspring is extremely low. The husband is relieved, rectifies his errors and wants to communicate this new information to his daughters. His wife, however, still doubts and even at the end of the session she sticks to her view that there was no genetic cause for her son's disease.

There are several processes of potential consensus formation in the pre-consultation period:

- in preceeding consultations with doctors and in hospitals (probably) consensus existed between clients and doctors;
- discussions about the symptoms and supposed diagnosis in the family;
- discussions about similar diseases in mass media;
- among clients: partial consensus about the preceeding events, disagreement on a perhaps correct diagnosis; disagreement on further measures to take (is genetic counseling necessary?)

During the consultation, consensus formation is required in several dimensions:

- between counselor and client: about a new diagnosis, about rectification of the error concerning the causes of the disease, about further measures;
- between counselor and female client: persisting disagreement on a correct diagnosis about the momentary and further measures to take: persisting disagreement on the interpretation of the preceeding events;
- between clients and family: consensus between husband and family about diagnosis and further measures, disagreement between wife and family on this.

Involved fields: counselor, clients, medical practice, public information, family.

Case 2: Age indication for prenatal diagnosis
A 37-year-old pregnant woman wants to have a prenatal diagnosis and decides to have a chorionic villi sampling (CVS). Neither she nor her gynecologist considers it necessary to have a consultation before the examination. The genetic counseling service and the gynecologist's department where the CVS will be done made an agreement that any CVS is preceded by a detailed consultation in the counseling service. Very reluctantly, some hours before the CVS, the client comes alone to the counseling session. During the conversation the counselee turns out to be very poorly informed about the medical implications and risks of this procedure and about alternatives. Finally, she becomes much inter-

ested in the information. The result of the prenatal diagnosis is a 47 XXY chromosomal disorder. When these findings are disclosed, both marital partners, a counselor and a social worker are present. Again, there is a detailed conversation about the findings, about the uncertainty of the diagnosis, about possibilities to check the diagnosis and about the Klinefelter's syndrome. From the very beginning the husband wishes that abortion be done immediately; the wife hardly shows any reaction. She keeps quiet and only answers when directly addressed. From her words one can conclude that although longing for the child dearly, she agrees with her husband. The counselor confirms that this diagnosis implies no indication for abortion.

In the pre-consultation period consensus existed in the following dimensions:

- between husband and wife: consensus about further measures to take (prenatal diagnosis);
- between wife and gynaecologist: consensus about the indication for prenatal diagnosis without any genetic consultation, the wife presumedly being sufficiently informed;
- between counselor and the department of gynecology: consensus about procedure in connection with CVS: consultation obligatory.

During the consultation interactions were the following:

- between counselor and wife: disagreement about the necessity of consultation, growing formation of consensus during the conversation.

Information about the findings led to consensus as well as dissensus:

- between counselor and clients: strong dissent between counselor and husband about the assessment of the findings; minor dissent between the counselor and the wife; general dissent among counselor and clients about the consequences (indication for abortion);
- between clients: consensus about abortion, dissent about the psychological problems the wife will be exposed to when having an abortion.

In the post-consultation period partial consensus existed:

- between clients: consensus about having an abortion: dissent with counselor will be avoided.

Involved fields: counselor, clients, human genetics, medical practice, professional and social context.

Case 3: Clinical Diagnosis: Myotonic Dystrophia

A 20-year-old woman belonging to a large family, where myotonic dystrophia has been transmitted for several generations, comes to a genetic counseling session. The client is very well informed about the symptoms by her own family. Her father, one sister and her child as well as many other members of the family are affected. Her father himself had a counseling session and informed his children thoroughly about the genetic background, the various forms of this disease and about prognosis. Clinical examinations confirm that she is a carrier of this gene. The woman does not consider the findings dramatic. Moreover, she is explicitly assured that these findings are not told to any other persons in order to avoid disadvantages when looking for a job or taking out insurance. The client confirms that she understands the situation but she is less worried than the counselor himself. Half a year later, she comes back reporting the following: she handed in her notice and looked for a corresponding job elsewhere. In the interview, she deliberately admitted being a carrier of the gene causing this disease. The potential employer consulted specialized literature and finally rejected her, because of her being a carrier of this disorder. The job was left vacant.

There are processes of consensus formation in the pre-consultation period:

- between client and her family: consensus about symptoms, about the genetic background, about evaluation of the disease, well-adjusted coping patterns, about the necessity of genetic counseling.

During the consultation the following framework of consensus and dissensus existed:

- between counselor and counselee: consensus about diagnosis, assessment of risk, prognosis, coping patterns, about required and practicable diagnosis for the client, about the client's family and offspring, consensus about not informing persons not involved, dissent about the possible social consequences.

In the post-consultation period consensus and dissensus interacted:

- the client informs her potential employer about her possibility of falling ill; dissent on the assessment of this disease.

Involved fields: counselor, clients, human genetics, professional and social context, family.

Case 4: Clinical Diagnosis: Hypochondroplasia:
A couple appears together with their two children; the younger child, a girl, suffers from hypochondroplasia; the parents, especially the mother, are very much worried about their daughter being undersized.

They have been informed that the recurrence risk of this disease is 100%; that with further children the reduced growth will be even worse. Genetic counseling seems to be their last hope to do something for their daughter. The mother is eager to be assured that the daughter will grow according to her wish. She is primarily interested to learn what they can do, to make her grow again. The clients have expected that in this consultation it will be checked, whether other members of the family are also affected with hypochondroplasia. The development of the consultation is strongly influenced by the mother insisting on her daughter being undersized. During the anamnesis of the family, conflicts between the partners arise and become more evident, when the husband learns that he is a carrier. The wife absolutely refuses to have further children although she would like to.

One year later, during the catamnesis it becomes evident that the marital conflicts have aggravated. The wife's insufficient knowledge concerning genetic transmission could not be corrected by the counselor's information. The husband, on the other hand, rectified his errors, remembers the counselor's information that he is a carrier; however, it becomes evident that he is not willing to accept the truth. As the wife keeps on misjudging the change of her daughter's size and the recurrence risk which, according to her, is 100%, she refuses to have further children in spite of her longing to have more.

There are processes of consensus formation during the pre-consultation period:
- among clients, doctors and clinics: consensus about the diagnosis (genetically caused hypochrondoplasia), consensus (or not ?) about the recurrence risks and prognosis;
- between the clients: dissent about guilt, consensus about the necessity of genetic counseling and possible therapies.

During the consultation several processes of consensus and dissensus manifested themselves:
- between counselor and clients: strong dissent about the expectations, aims, counsels and risks of recurrence;
- between counselor and husband: consensus about social and psychological assessment of the disorder, dissent about genetic findings (that he is a carrier);
- between counselor and female client: consensus about genetic findings of her husband;
- dissent between husband and wife about social and psychological

assessment of the disorder, persisting dissent about the interpretation of the genetic background and its consequences for the partnership.

Involved fields: counselor, clients, human genetics, medical practice, professional and social context, family.

II. THE GENETIC COUNSELING SESSION, EXPLICIT AND IMPLICIT PROCESSES OF CONSENSUS FORMATION

In the first section, the complex system of the process of genetic counseling was schematically illustrated and explained by four cases. In the following, we discuss the development of consensus formation in the counseling session itself. We will concentrate on the conversation between counselor and counselee and disregard the pre-consultation-process which might have influenced the conversation. Information on the contents of the counseling session and the (medical) history of the persons involved will determine the counselor's and counselee's expectations and attitudes. It will influence a counseling session in advance. The analysis of genetic counseling sessions illustrates that the structure of the conversation as well as its logical development have typical characteristics. The structural schemes of counseling processes differ according to the center of interest of the person engaged in analyzing genetic counseling sessions. They might primarily focus on the formal sequence of actions, or on the presentation and elaboration of specific problems ([6,8,9]). As the course of a counseling session always depends on the character of the counselee(s), the counselor's individual working style and the problems of the case, very individual and subtle schemes can be described for each counseling session [9]. Schemes of different consultations therefore might be very diverse. In spite of that complexity and diversity all counseling sessions can at least be subdivided into the following main phases: (a) introduction and opening of the conversation (b) exchange of information and discussion of the problem (c) conclusion and summary. During any of these phases explicit or/and implicit processes of consensus formation might happen between counselor and counselees. These processes might appear more or less important to the persons involved; they might be successful or not.

(a) Introduction and opening of the conversation

Generally, the counselor and his client(s) meet for the first time in the waiting or consultation room, where they are introduced to each other. This is followed by the invitation to have a seat and then they indulge in small talk (e.g., about the arrival, about the site, the atmosphere and so on). This phase is a kind of warming-up phase influencing considerably the further development of the inter-human communicative process. In these first minutes the later relationship between counselor and patient often is already decisively influenced by a few words alone accompanied by non-verbal elements like welcoming gestures, first impression of outward appearance, seating order, etc.. Spontaneous feelings of sympathy or antipathy are evoked. Correspondences or differences in the idiosyncrasies and lifestyles are silently observed. In this phase, the unbalanced relationship between counselor and counselee, which is said to be a typical phenomenon of genetic counseling sessions, becomes most evident. There is a difference in knowledge and in attitude and a different degree of being involved [8]. In the next sequence, following this warming-up phase, the reason for the counseling session and the specific "problem" of the client must be named. After a first vague presentation of the problem (usually done by the patient; one might already establish a typology of different problem-presentations here) the problem must be defined. In an interactive process between counselor and counselee, the question arises, which problem the session will and can focus on. Case 4 illustrates very well that during the process of problem-presentation and the interactive problem-definition (the mother is merely interested in new therapeutic measures for the daughter, the counselor focuses on the risk of recurrence with further children) fundamental differences in the evaluation of facts might arise between the interlocutors. These differences might counteract further understanding and consensus formation from the very beginning of the consultation.

(b) Exchange of information and discussion of the problem

This first understanding between counselor and counselee about the motive and the "problem" of the genetic counseling is generally followed by a phase of intensive exchange of information. The counselor tries to build up a coherent picture from the information given to make a precise diagnosis and to define predictable calculable risks. He will try to present the disease adequately, to explain its possible forms, and to inform the client in a comprehensive way about possible genetic and other risks. The question of objective or subjective risk assessment plays a decisive role in this phase of the counseling session. Sometimes, the counselor is confronted with the problem that he cannot define the risk as precisely as the client expects him to do. This is specially true for obscure ways of transmission; rare diseases or even today vaguely calculable risks (e.g., environmental risks, questionable exposure to radiation, etc.). Added to this problem of being unable in many cases to indicate the exact percentage of the risk, comes another problem, namely, the fact that individual clients understand and interpret risk rates very differently: when you tell a patient that there will be a risk of 5 – 10% of severe malformation this might appear absolutely unacceptable to her and suggest abortion in early pregnancy. To another client, the disclosure that in spite of a specific personal genetic risk the child, she is expecting, has a chance of 90 – 95% of not suffering from a genetically caused disease might calm her very much. The subjective interpretation and evaluation of risks are strongly influenced by the client's social class and cultural surroundings, the individual value system and lifestyle.

Especially when discussing risk assessment it becomes evident that to the counselee not the simple explanation of a rather precisely definable risk, but the subjective evaluation and interpretation of this risk is of greater importance. Therefore, the counselor should extend the conversation by coming back to the following topics: the counselee's personal experiences, his/her knowledge about the disease, his/her biographic background, intellect, social surroundings, system of values and ideology. All these elements help to explain his or her personal interpretation of the risk which possibly differs considerably from the counselor's [7]. However, one can find this hermeneutic circle of understanding and interest [5] on both sides. It also influences the counselor's speech and behavior: his presentation of possible risks and his way of explaining these risks (he might quote manuals or authorities, show rather drastic photos, tell other case histories) depends on his personal values and his convictions. Although the counselor's professional ideology is influenced by a professional, medical understanding of counseling, this does not imply that the process of genetic counseling can and should be a totally value-free process of information. On the contrary, such a subtle discussion about different ideologies, values, aims and consequently, subjective situative interpretations of objective findings, might be decisive characteristics of a successful counseling session. It goes without saying that the thoroughness and candor in the integration of value-related aspects and anticipation of events in the dialogue of genetic counseling depend on the counselor's character and on his working style. Only in a rather open atmosphere, characterized by mutual empathy, is it possible to talk adequately about feelings and convictions accompanying vital decisions. Today, the majority of the genetic counselors in Germany feel obliged to promote an autonomous decision of the clients, and prefer a non-directive way of counseling ([1,2,11,17]). This does not suggest that the counselor should not utter his personal convictions or personal arguments. It might be of some help to the client to learn that the counselor's values and convictions might differ from the client's. Thus, the degree of transparency of the individual system of values ensuring individual decisions becomes an essential criterion of the candor of the dialogue about diagnosis, risk rates, prognosis and, consequently, about attitudes and decisions.

(c) Conclusion and summary

In the concluding phase of the session, the findings and results, as well as the opinions, hopes and fears are summarized. As the moment has come to take further decisions concerning the following steps (e.g. amniocentesis – yes or no?; involving relatives in the explanation of the diagnosis etc.) it can be clearly seen whether a general consensus – or a partial consensus on specific topics – was obtained in the consultation. When actually formulating the options of decision and attitude, possible collisions between the counselor's and counselee's values and aims become evident and are explicitly named (see case 2). At this phase of the consultation process a re-beginning of the counseling dialogue might hint at a not yet successful process of mutual understanding and at a deficient consensus formation. Even the often inevitable question coming up at the end of the consultation: "Doctor, what would you do in my place"? is ambiguous and very telling [1]. Considering its function in the process of the conversation, it might imply that the client, confronted with data in profusion (e.g., different rates of risk, different degrees of manifestation of genetic disorders, etc.) feels totally incapable of making a decision. The consultation left him rather helpless and without orientation, and the question expresses his need of further counseling. On the other hand, this question might show that after there was consensus between the counselor and the counselee about a non-directive style of communication, the client finally wants to learn the counselor's personal opinion. This suggests that the counseling session has been positively influenced by the non-conducting way of counseling. Even if this is the case, it is doubtful whether the counselor should give an answer based on his own system of values and philosophy, or rather consider the counselee's situation and adopt his or her philosophy of life. Last but not least, one should always bear in mind that a conclusively formulated decision of what to do must not only consider the counselor's and client's personal interests, but even the interests of the further persons involved, e.g., the unborn child, relatives. This draws our attention again to the fields of interaction displayed in Figure 2.

Figure 2. Process of interaction in genetic counselling

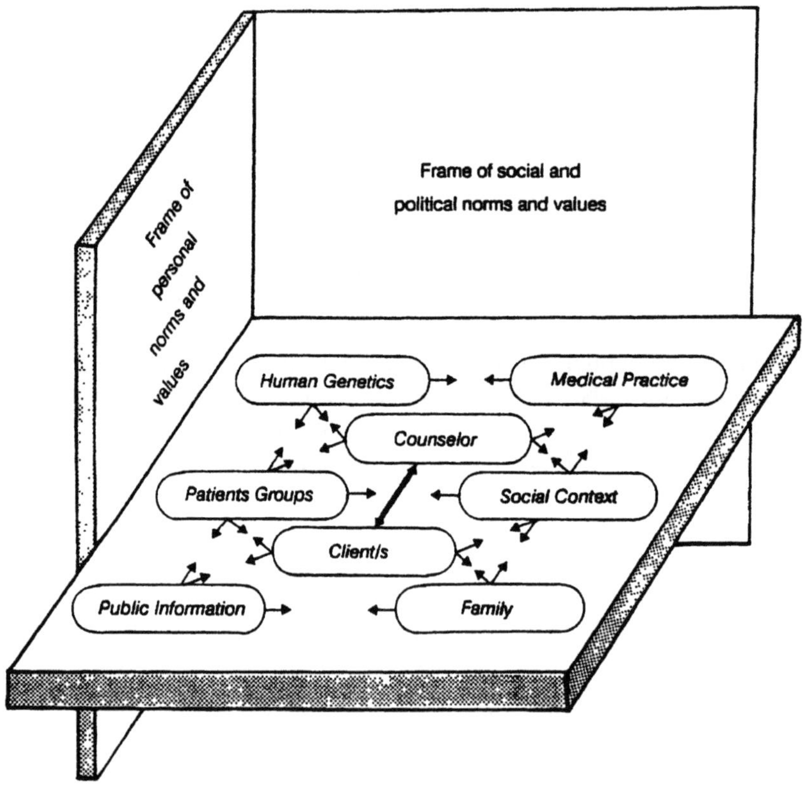

The concluding report containing information about the findings, that follows most genetic counseling sessions has two aims. On the one hand, it serves the counselee as written documentation and on the other hand, the results of the counseling process are accessible to other persons involved, e.g., the referring doctors, the gynecologist's department. Before the drafting of the concluding report, sometimes a lot of additional feedbacks, examinations and conversations have to be made. However, in this phase prompt reaction is required. So agreements on how to proceed have to be made by phone, by brief notes, and must often be transmitted to the appropriate authorities before the final report is finished [10].

III. CONCLUDING REMARKS

The most important aim of the communication process in genetic counseling is to help the clients make a decision they are responsible for themselves. A look at the complex field of interaction in genetic counseling reveals that this process of communication can also be described as a sequence of decisions: decisions already taken or to be taken before the counseling session. In the pre-consultation period, the clients make decisions about the acceptance of possible factors, about their assessment, about anticipations and prognosis, and about measures to take. The clients' anticipation of events, their reasoning and thinking in scenarios are very often influenced by emotions, especially fears. This sequence of decisions goes on in the counseling sessions. The counselor and his client(s) have to agree on the manner in which the counseling will take place. The contents, prerequisites and consequences of decisions taken or to be taken are discussed, varied, or modified. Consensus is obtained; dissent about expectations, risks and their assessment about causes and consequences are observed. More decisions will be taken, about contents and facts (whether they are acceptable or rejectable) or about communication and consensus formation (was it a good/bad conversation; was it successful, was it a failure?). This sequence of decisionmaking goes on in the final phase of the session: the relevant factors are assessed so that the clients might combine them in the right way, before their final decision. This is the most critical core of genetic counseling where the main problems of consensus formation between counselor and clients arise. The crucial problem for the clients is the fact that they can only opt for a decision within a binary system: prenatal diagnosis – yes or no?; Acceptance of a handicapped child – yes or no?; Abortion – yes or no?

However, the factors accompanying this process of decision making are multidimensional. They differ according to their contents, meanings, implications and assessments. They are influenced by emotions (e.g., expectations, fears). This problem, that after complex and thorough reflections and after the discussion of all imaginable scenarios it is up to the clients to make a final (and often an irreversible) decision, generally characterizes the final phase of the counseling session. Moreover, even in this phase the unbalanced relationship between the counselor and the clients becomes evident. Although the counselor's empathy might be great, it cannot counterbalance the counselor's and counselee's different degree of being concerned: the client himself is the person being confronted with the consequences arising from his decisions. What are the criteria for a good or bad counseling session? When are we entitled to call it a success or failure? First of all, we must realise that the answer depends on the perspective. We must ask ourselves when and under which conditions the clients will experience a consultation as successful, good or helpful. Which criteria does the counselor use, when he approves of the consultation or is dissatisfied. We have to consider and to accept that the counselor's and client's criteria are not always identical. Consequently, the counselor's and the client's assessment of the consultation will differ. We might assume that the client's criteria vary from person to person. This is also true for the counselor's criteria. Moreover, the counselor's capability and willingness to define his expectations, values and criteria, which will influence the course of the consultation and its assessment, will vary. Therefore, in the professional training of genetic counselors one should attach more importance to the reflection about one's own norms and values. The final assessment of the genetic counseling session (i.e., whether it was successful or not?) will depend considerably on the counselor's command of the above mentioned faculties. It would be inappropriate to assume that consensus is achieved only if the client's final decision corresponds with the counselor's ideology and system of values. In genetic counseling, the process of consensus formation cannot be located in one dimension, but it depends on a rather complex interpersonal system.

Department of Psychotherapy
University of Ulm, Ulm, Germany

BIBLIOGRAPHY

1. Beck-Gernsheim, E.: 1995, 'Genetische Beratung im Spannungsfeld zwischen Klientenwünschen und gesellschaftlichem Erwartungsdruck', in Beck-Gernsheim, E, (ed.) *Welche Gesundheit wollen wir? Dilemmata des medizintechnischen Fortschritts*, Suhrkamp, Frankfurt, pp. 111-138.
2. Bundesminister für Forschung und Technologie (ed.): 1991, *Die Erforschung des menschlichen Genoms: Ethische und soziale Aspekte*, Campus, Frankfurt, New York.
3. Fässler-Trost, A.: 1989, *Über das "Ratgeben" in der genetischen Beratung. Untersuchungen zum verbalen Interaktionsverhalten des Beraters und dessen Wertorientierungen*, PSZ-Verlag, Ulm.
4. Fraser, F.C.: 1974, 'Genetic Counseling', *American Journal of Human Genetics* **26**, 636-659.
5. Habermas, J.: 1968, *Erkenntnis und Interesse*, Suhrkamp, Frankfurt/Main.
6. Kessler, S. (ed.): 1979, *Genetic Counseling. Psychological Dimensions*, Academic Press, New York, San Francisco, London.
7. Kessler, S. (ed): 'Current psychological issues in genetic counseling', in Endres, M. (ed.): *Workshop on Psychological Aspects of Genetic Counseling 1989, Journal of Psychosomatic Obstetrics and Gynaecology* **11**, Supplement 11.
8. Nothdurft, W.: 1984, *"...Äh folgendes Problem äh..."*, Narr, Tübingen.
9. Reif, M. and Baitsch, H.: 1986, *Genetische Beratung – Hilfestellung für eine selbstverantwortliche Entscheidung*, Springer, Berlin, Heidelberg, New York, London, Paris, Tokyo.
10. Reiter, J. and Theile, U.: 1985, *Genetik und Moral: Beiträge zu einer Genetik des Ungeborenen*, Matthias-Grünewald, Mainz.
11. Schöne-Seifert, B. and Krüger, L.: 1993, 'Humangenetik heute: umstrittene ethische Grundfragen', in Schöne-Seifert, B. and Krüger, L. (eds.), *Humangenetik – Ethische Probleme der Beratung, Diagnostik und Forschung*, Fischer, Stuttgart, Jena, New York, pp. 253-289.
12. Schröder-Kurth, T.M.: 1989, *Medizinische Genetik in der Bundesrepublik Deutschland. Eine Bestandsaufnahme mit politischen, ärztlichen und ethischen Konzepten, Stellungnahmen von Patientengruppen*, Schweitzer, Neuwied/Frankfurt.
13. Sponholz, G.: 1990, *Untersuchungen zur formalen Struktur des genetischen Beratungsgesprächs. Computerunterstützte Inhaltsanalyse an 37 Beratungsgesprächen*, Dissertation, Universität Ulm.
14. Wertz, D.C. and Fletcher, J.C. (eds.): 1989, *Ethics and Human Genetics. A Cross-Cultural Perspective*, Springer, Berlin, Heidelberg, New York, London, Paris, Tokyo.
15. Wolff, G.: 1981, 'Genetische Beratung als Kommunikationsprozeß I. Ein Beitrag zur Praxis der genetischen Beratung', *Sozialpädiatrie* **3** (1), 35-37.
16. Wolff, G.: 1981, 'Genetische Beratung als Kommunikationsprozeß II. Ein Beitrag zur Praxis der genetischen Beratung', *Sozialpädiatrie* **3** (2), 91-93.
17. Wolff, G.: 1989, 'Die ethischen Konflikte durch die humangenetische Diagnostik', *Ethik in der Medizin* **1**, 184-194.

SANDRO SPINSANTI

OBTAINING CONSENT FROM THE FAMILY: A HORIZON FOR CLINICAL ETHICS*

I. INTRODUCTION

To begin, I would like to report something personal – not about obtaining patient consent in clinical practice but about what can be called "physician consent." In clinical ethics we are faced with the often difficult task of procuring the doctor's "consent" to cooperate with those who do not belong to the medical corpus but who attend to what occurs in medical practice. I believe that it is an experience common to many of us who are "outsiders," such as philosophers, bioethicists, theologians, and jurists, that the physician's first reaction with regard to our intervention in the medical field is one of diffidence and skepticism. As we are laypersons from the doctors' perspective, we must overcome a subtle barrier which can be seen in small backward movements of the head, a raised eyebrow accompanied by a questioning look, a barely detectable twitch of the mouth followed by a slightly ironic smile which says: What makes you think you know medicine? It is an invisible barrier, but one that is nevertheless real. Sometime ago I mentioned to a physician friend that I was planning to participate in a conference on the topic of gaining consent in clinical practice. Immediately I saw that skeptical expression (which I now know well) appear on her face. At the same time, her mouth was formulating the question: Consent, what is that? I tried to explain briefly what I meant. It was not so difficult to overcome the barrier – at least that time. With a smile which signified insight, she told me that the week before there had been a case in her hospital in which it was necessary to obtain consent from the family for one of the patients. It was difficult; but, after a long discussion, she had succeeded. Then she related the following story.

II. CASE STUDY

Mr. M, age 70, was hospitalized in our unit, a clinic specializing in lung diseases. A bronchoscopy showed a pulmonary tumor in an advanced

H.A.M.J. ten Have and H.-M. Sass (eds.), Consensus Formation in Healthcare Ethics, 209–217

stage, having already reached the two bronchi and the trachea. Severe dyspnea was slightly and temporarily relieved by an immediate laser resection.

Four days after Mr. M's admission, I was told by my colleague of the patient's condition: he was getting worse and death was imminent. I called his family which included his wife, two sons about 40 years of age and their respective wives, and a daughter of about 30. The elder son spoke to me first and asked to know the details of his father's condition. I said to him that, as he already knew, his father was affected by a progressive malignant tumor. He objected strongly to this statement, claiming that he didn't know, that no one had informed him. When I checked with the nurse, however, she told me that she had been present at the discussion during which the two sons had been informed. I tried to explain the situation to the son once again. I told him that his father was affected by a tumor of the bronchi in an advanced stage, that there was no possibility of treatment which would benefit him, and that death would probably occur in a few hours. He seemed astonished. He repeated several times that this was not possible. Just the week before, he said, his father had been sitting in the garden. It might well be a malignant tumor, but such a rapid death was not possible. He pleaded with me to do everything I could for his father. Medicine today has made such progress, he said, that it must be possible to do more. Couldn't we use laser treatments again? Or transfer him to the ICU and let him breathe by a respirator (ventilator)? He insisted that every known therapy be tried for his father. The other son agreed with his brother that aggressive treatment ought to be tried. Even a talk with the chief physician failed to dissuade them. The medical team discussed whether there might still be some benefit from laser treatment and came to the conclusion that the burdens would outweigh the benefits. Even if it were successful, it would extend his life for a very brief period only; he would suffer from extreme dypsnea and would still not be in a condition to be discharged from the hospital. We agreed that none of us would have consented if this were our own father.

Meanwhile, the global respiratory insufficiency increased and Mr. M. went into a coma. We discussed the situation once more at the patient's bedside. As simply and clearly as possible, the doctors told the family what the opinion of the medical team had been. The chief physician explained that the surgery required anesthesia which meant a high level of risk for a moribund patient. For that reason it was not recommended. He then turned to the nurse and asked for her opinion. She replied spontaneously that she would let the family decide. The chief physician once again outlined the two scenarios: on the one hand the patient would be in a state of drowsiness, very likely without pain, close to death, and surrounded by his extended family (many other members of the family had begun to arrive). On the other hand, the patient would endure an aggressive procedure during which he might die on the bronchoscopy table; if he were to survive, it would be with a respiratory insufficiency and his condition would continue to deteriorate. There was no possibility that he would be able to return home. The physicians' opinion was that the best decision would be to do nothing. However, they would treat aggressively if the family insisted that they pursue this course of action. The family talked among themselves for about half an hour, after which they informed us that they did not want any further intervention. The wife and daughter told us clearly that they agreed with us; it was the sons who found it difficult to accept the choice not to do something. Mr. M. died an hour later. The family thanked us and left the hospital in great sorrow, but without any anger.

III. THE PLACE OF THE FAMILY IN BIOETHICAL GUIDELINES

From the case study we see that, in good medicine, the physician usually considers the family. Do we need to take this relational network of the patient into account when developing ethical guidelines for the physician? In order to answer this question, let us first take a look at the documents which have traditionally directed the conduct of European physicians on this matter. On January 6, 1987, a *European Guide of Medical Ethics* was approved at the Conference of the EEC Medical Order.[1] If we look through the document, we do not find any references to the patient's family. The guidelines presume, theoretically, an absolutely individual conception of the physician-patient relationship. The primary notions which are to ethically guide the decisionmaking process are patient benefit and physician autonomy. This physician autonomy removes him or her from the realm of social regulation regarding medical practice. As a result, problems of cost have no relevance when looking at the matter of consent from this perspective only. On the other hand, the strong emphasis put on beneficence implies an anthropological concept which considers the person as an individual instead of in relational terms. The type of physician-patient relationship which inspired the *European Guide of*

Medical Ethics is described as a contract between two autonomous individuals who can agree only by compromise. In contrast to this approach is the relational model in which a human being constitutes himself or herself, from birth to death, in a network of relations of which the family is the symbolic referent. The first model, as instantiated in the *European Guide*, is the most successful attempt to integrate the principle of autonomy as developed in the American bioethical milieu.[2] The second has been a *de facto* practice in the Mediterranean cultural environment throughout history.

It may seem as if, in the Latin context, the family occupies a more important position than the individual in medical decisionmaking. The professional code of Italian physicians contains explicit mentioning of the family in the chapter which guides the physician-patient relationship. This is brought to the forefront in the situation which, for a doctor in a Latin culture, constitutes the ethical dilemma par excellence: Should an unfavorable diagnosis or prognosis be communicated to a patient? On this question, the previous Italian medical professional medical code, elaborated by the Italian Medical Association in 1978, stated in Article 30: A serious or unfavorable prognosis may be kept hidden from the patient, but not from his or her family. The more recent version of this professional code, approved in 1989, states in Article 39: The doctor may take the opportunity, after evaluation, to minimize or hide from the patient the seriousness of an unfavorable prognosis; but it must be communicated to the intimate others. In any case, the wishes of the patient, expressed freely, must represent a determining element for the doctor, and his or her own behavior shall be inspired by the same.

From the comparison of these two versions, formulated ten years apart, some interesting observations arise. First of all, the word 'family' has been replaced by 'intimate others.' Evidently even in Italy the traditional image of the family in which the existing relations substantially reflect the close support system as legally registered has given way to include more fluid and irregular situations. With the new formulation, it is possible to give the same value to those living with the patient but not formally related as is the 'family' of the patient. This would, in fact, include anyone who has a significant relationship with the patient. A second element, which is substantially more important, is the section which refers to the patient's wishes as an indicative standard. This is a new element in Italian medical professional ethics which implies the overcoming of two traditional attitudes: medical paternalism, i.e., that the doctor knows what is best for his patient, and 'familism', which places the preferences of an individual second to that of the related group. With this concession to the principle of autonomy in the Italian physician's code of ethics, there still remains culturally a relational orientation which makes it necessary for the doctor to take into consideration not only the wishes of the patient as an individual but also those of his or her family or intimate others, if possible. As a result of this, we may perhaps imagine two different models of consent: one is centered on the individual person and is intended to protect his or her autonomy; the other is oriented towards the person in his or her relational dimension and is intended to promote greater caring.

IV. ETHICS OF CARE AND ETHICS OF JUSTICE

As often happens in ethical thought, our first effort must aim at avoiding any form of dualism which tends to place good and evil on opposite sides. We must also be careful not to surrender to the temptation to value one model more highly because it is more "modern" while devaluing the other because it is a continuation of a tradition which is "destined" to give way to a more progressive conception. Certainly today it has become a trend in bioethical thought to denounce the limits of an "ethics of justice" which does not integrate the "ethics of care" as first presented by Carol Gilligan [1]. This ethics of justice, the final aim of which is to instantiate the principle of autonomy, does not lead to the strengthening of interpersonal ties in family and community; it perpetuates fragile relations based on contractarian reasons. The ethics of care, however, is based on the assumption that persons are dynamically interconnected and that each situation requires a joint evaluation of the interreactions. Therefore, fairness in the consent process is not enough; consent must be a factor in preserving the relations. As with any moral judgment, the process of forming consent cannot be reached by a solely *rational* method; it must also include a *relational* method.

Whatever credit might be given to the ethics of care as a corrective of the unilateral aspect of the ethics of justice and the application of autonomy, I believe we must never try to discredit it. Autonomy is an important value and it must be promoted, especially in those cultures in which medical ethics still tends to be dominated by paternalism. The promotion of autonomy is not limited to the countering of medical paternalism; it is also an effort to protect the patient from intrusions of the family. We cannot afford to lose sight of the individual's best interest. There is a need to be sensitive to the chance that the consent which the physician obtains from the family might be the result of a collusion, knowingly or unknowingly, which will work to the patient's disadvantage. When speaking of collusion we must not only think of situations in which the family agrees with the doctor about one or another therapeutic strategy which will affect the survival of the patient; we must also consider, when there is agreement, what the material interests might be, such as a will or inheritance. These latter are cases which involve the civil code and the courts rather than ethics alone. There are, however, situations in which the physician's collusion with the family has much more subtle aspects. A convincing example of such a situation is offered in a detailed anthropological study carried out by Deborah Gordon, an American anthropologist, on the communication of the diagnosis of breast cancer to women in Italy, particularly in Florence [2]. Even in the case of breast cancer, Deborah Gordon found confirmation of what had already emerged from other research involving patients with neoplasms: the diagnosis was not generally communicated to the patient but to the family; even when it was, it could not be truly called a communication because it was full of euphemisms and reticence and even, at times, intentional lies. In these cases it was found that invariably the family chose, in agreement with the physician, to withhold information from the patient in order to protect the patient from shock. It was always considered to be "for her own good."

V. THE BENEFICENCE MODEL OF CARING FROM A JUSTICE PERSPECTIVE

Anthropological studies have made a contribution by deeply analyzing the cultural mechanisms underlying choices to withhold information, choices which *prima facie* appear to be ethically motivated. In the Italian culture, where there is still a strong association of cancer with death, suffering and lack of hope, the non-communication of the diagnosis is equivalent to a mechanism which is designed to keep the "condemned" patient "with us" in the social world while leaving death and suffering in the "other" distant world. What dominates is the social reality, while informing a patient of the diagnosis is equivalent to social death. Deborah Gordon observes:

> This non-disclosure augments the experience of a divided world for many patients. In many ways cancer is an illness of divisions, of disunity, of otherness. The illness itself is often lived as 'other', the person with cancer the same. Both medical and popular accounts present a battle between the 'good and the 'bad', the 'benign' and the 'malign', reasserting the orderly dichotomous understanding of the world that cancer in fact defies([2], p. 292).

If this interpretation may be confirmed – if, in other words, the non-communication of the diagnosis may be seen as the elimination from the fabric of the community of the "other" and is intended as a symbol of death which threatens the social body – then what is apparently a caring model in which the doctor seeks the consent of the family in order to protect the patient from the anguish of death, shows all of its ambiguous nature. Such an alliance becomes a collusion which will cost the patient great anguish and even extreme isolation. Such an "ethics of care" needs a good injection of autonomy as a corrective agent. And the physician, instead of seeking the consent of the family should try to face their disapproval in order to protect the patient's right to manage his or her own life. Obtaining the consent of the family, instead of always being a guarantee of good medicine practiced with a high ethical profile, may become instead a form of prevarication.

VI. CONCLUSION

We cannot state that the involvement of the family in the obtaining of consent is an optional extra, as bioethical models centered on autonomy would lead us to imagine. In fact, in most of the protocols which are inspired by these models there is no place for family involvement. As noted, however, we also cannot state that the consent of the family is always the most beneficent choice. Perhaps it is closer to reality when we state that the two ideal models must be integrated. Edmund Pellegrino and David Thomasma have offered a basis for this integration in a principle they term 'beneficence-in-trust.'

> But healing, as we define it, is a form of assistance in making the patient whole again by working through his or her body. If the values of patient welfare and patient autonomy remain in conflict, then authentic healing cannot take place. A physician, therefore, must become both a

> moderate autonomist and a moderate welfarist. This can be enhanced in a beneficence model such as the one we suggest ([5], p. 35).

The integration of beneficence and autonomy is not an easy process.[3] Scientific medicine may believe that informed consent is superfluous, claiming that in the physician's experience with sick patients he or she has learned more about the sphere of human relations than other interested persons. Bioethicists have had greater success in providing for informed consent, and the beneficence-in-trust model is a unique theoretical approach to the integration of beneficence and autonomy. Most models, however, are not built on a relational basis and do not offer even a theoretical framework that might take the family into account. After having looked at some of the considerations involved in developing such a model, we can conclude that the inclusion of the family in the informed consent process is both caring and just. It is caring in that it reminds us of the fact that, without consideration of the human being as a relational being, without making decisions concerning the patient in the context of a network of interpersonal relations, respect for the person is not honored. It is just in that it includes those who are part of the patient's *Gestalt*, those who form the network in which the patient lives and finds an identity. Without such a consideration, the patient is not seen and treated in context. It is only possible to practice "good medicine" in the clinical sense when the patient is understood from this wider perspective. And without good clinical medicine, there can be no ethical practice of medicine.

Istituto Giano
Rome, Italy

NOTES

* Translated and edited from Italian by Patricia Mazzarella.

1 The Guide makes two types of recommendation: that relating to general duties and that relating to specific rules governing biomedical innovation encountered in such activities as human research and genetic screening. In the history of medical ethics, the first category was developed from the Hippocratic Oath and subsequent tradition. The second category, however, has been only vaguely defined without specific reference to the directives of the CIOMS and to the International Code of Medical Ethics formulated by the World Medical Association in London, 1949 and Sidney, 1966.

[2] Recently, some American bioethicists have expressed their concern about the prevalent ethic of patient autonomy which ignores family interests in medical treatment decisions. See [4].

[3] For a valuable contribution to the integration of the models of autonomy and beneficence which arises from the "Mediterranean tradition" of sensivity to the family, see [3].

BIBLIOGRAPHY

1. Gilligan, C.: 1982, *In a Different Voice: Psychological Theory and Women's Development*, Harvard University Press, Cambridge, Massachusetts.
2. Gordon, D.: 1990, 'Embodying Illness, Embodying Cancer', *Culture, Medicine and Psychiatry* **14**, 275-297.
3. Gracia, D.: 1989, 'Una barriera tra stato e individuo', *Famiglia Oggi* (no. 37), 88-90.
4. Hardwig, J.: 1990, 'What about the Family?', *Hastings Center Report* **20** (2), 162-171.
5. Pellegrino, E. and Thomasma, D.: 1988, *For the Patient's Good: The Restoration of Beneficence in Health Care*, Oxford University Press, London.

CHRIS HACKLER

CONSENSUS AND FUTILITY OF TREATMENT: SOME POLICY SUGGESTIONS

The current emphasis on the rights of patients and families to participate in treatment decisions is a welcome change from the paternalism that has characterized medical practice through most of its history. Traditional codes and treatises on medical ethics emphasized the responsibilities of physicians to benefit their patients and to avoid harming or exploiting them. Physicians were assumed to be in the best position to make medical decisions and were required to exercise their authority in the best interests of their patients. Only recently have codes and books on medical ethics asserted the right of patients to participate in decisions about their care, primarily in those societies historically committed to the importance of individual liberty. Consensus between physician and patient on treatment decisions is still a relatively new concept.

The primacy of patients' rights has gradually become embodied in the policies of hospitals and other healthcare institutions in the United States. Physicians must obtain informed consent for any treatment that is not routine, and patients who are capable of making responsible decisions may reject any treatment they do not want. If a patient is not capable of deliberation and choice, then family members or other suitable surrogates may choose on behalf of the patient. The primary role of the surrogate is to extend patient self-determination by deciding what the patient would choose if capable of choice. If there is little evidence for what the patient would choose, or if the patient is a child or an adult who has never achieved decision-making capacity, then the surrogate's proper role is to help determine the best course of treatment for the patient. Family members should be in the best position to know what the patient would want, and they may be presumed to want what is best for the patient. In addition the family is generally considered to be a social unit with considerable authority over its own affairs. The participation of family members is especially important when a life-or-death treatment decision must be made. For this reason hospitals generally require agreement between the attending physician and the patient's next-of-kin, guardian, or closest family member, before life-prolonging procedures can be withheld or withdrawn. Policies mandating such a consensus are appropriate and

H.A.M.J. ten Have and H.-M. Sass (eds.), Consensus Formation in Healthcare Ethics, 219–228

desirable as a protection of patient values and interests, but a policy that grants unlimited control to family members or admits of no exception can produce serious problems for both patients and physicians. It can subject patients to pointless suffering and require physicians to violate their Hippocratic duty to protect the patient from harm. In extreme cases of this sort, it should be possible for a physician to refuse to comply with family wishes and to withdraw life-prolonging technology. As a possible example of such a case, please consider the following.

I. THE CASE OF BABY RENA [13]

An 18-month-old baby named Rena, abandoned by her mother at birth, had been in the intensive care unit for several weeks. She was dying of AIDS and a heart disease and could breathe only with the assistance of a respirator. She had little hearing or sight. She was fed through intravenous tubes and had chronic diarrhea. Her pain was so great that her physician usually kept her sedated. When she was moved for any reason her blood pressure increased dramatically and she cried – or at least she shed tears; she could make no sounds because of the tube in her throat.

The baby's life could be prolonged for weeks or even months by continued aggressive treatment, but it was certain that she would die and that her remaining life would be filled with little but misery. Her physician did not want to continue the treatment, which he viewed as pointless torment. He tried to explain to the baby's foster parents that she was suffering and that there was no hope of recovery. The foster parents were evangelical Christians who believed that only God could make the decision to allow a person to die. They believed that God told them to rear the child "in the nurture and admonition of God's word so that she would be a testimony to the body of Christ." They seemed to view the situation as a test of their faith in God's power to heal and of their submission to divine dominion over matters of life and death. They were hoping for a miracle. They spoke of treating the spiritual aspect of the baby's condition and fighting the "spirits of infirmity" that possessed her body. When physicians explained that she was in great physical pain, they replied that it was God's will. They insisted that everything be done to prolong life as long as possible.

The physician and all other members of the healthcare team felt strongly that it was wrong to prolong the suffering of the child. They wanted to cut short the misery by ceasing respiratory support and allowing her to die. The physician was willing to issue an order to this effect, but hospital regulations would not allow it. Like most other hospitals, this one required the approval of a parent or legal guardian before life-sustaining measures could be discontinued. The physician felt forced by this policy to practice bad and even unethical medicine – in effect, to torture his patient.

II. CONSENSUS AND HOSPITAL POLICY

Cases like Baby Rena's occur with increasing frequency in hospitals in the United States. The parents' or family's motivation is not always religious; it may be guilt for previously ignoring or mistreating the patient, inability to cope with death, or a number of other factors. In many cases the family members have good intentions and are suffering deeply themselves. While such good intentions and suffering should be respected, they do not in themselves entitle family members to force agonizing procedures on helpless patients. Hospitals should reconsider any policy that in effect creates an unrestricted right to demand such treatment. They should allow physicians to discontinue life-support without family consent in something like the following set of circumstances: (1) the patient is unable to consent, (2) the uncontested medical opinion is that the burdens of continued existence clearly outweigh the benefits and there is no reasonable chance of recovery, (3) the surrogate decision-maker does not provide an appropriate kind of reason for refusing to consent, and (4) the physician has made sufficient attempt to communicate with the surrogate and to mediate the disagreement.

A closer look at each of these conditions may be useful. (1) The first condition should be obvious: the patient has permanently lost the capacity to make decisions about treatment. Patients who retain this capacity have the right to decide for themselves whether to accept or reject treatment, and if there is reasonable hope that a loss of competence is only temporary, treatment may be continued until the patient recovers the ability to decide for himself. (2) If responsible medical opinion is divided over the relative benefit to the patient of continued treatment, then deference should be shown to the family's wishes. Surrogates must have the right to choose among legitimate medical options, or the whole point of surrogate decisionmaking will be lost. They must be able to weigh the physicians' assessments of the patient's present quality of life and chances of improvement and choose the course of treatment that seems to them best. It is only when there is a consensus of medical opinion that the patient is suffering with no reasonable hope of improvement that action might be taken without physician-family consensus.

Condition (3) is that the surrogate will not provide an appropriate kind of reason for insisting on continued treatment. The appropriateness of surrogate reasoning should not be simply a matter of subjective physician judgement. There is a broad consensus that the role of surrogates is two-fold. First, they should try to determine what the patient would choose if she were able to do so. Advance directives generally provide the most reliable substantiation of patient preferences, but less formal kinds of evidence may also be considered: oral instructions, letters, conversations, previous choices, or patterns of behavior that illustrate relevant values. For some patients there will be insufficient evidence to reach a reliable conclusion. For others there will be no such history at all – babies, for example or individuals who have been severely mentally retarded from birth. For patients such as these, surrogates must consider the second appropriate issue: What is in the patient's best interest? While there certainly is some room for disagreement about what is best for a given patient, it is not entirely a subjective matter. The President's Commission for the Study of Ethical Problems in Medicine and Biomedical and Behavioral Research is surely correct in expecting surrogates to follow general standards of reasonableness by considering such things as relief of suffering, restoration of functioning, and the quality and extent of life that is prolonged ([8], p. 180). If surrogates steadfastly refuse to confront either of these two concerns – what the patient would choose or what is best for the patient – then they are not providing reasons that are appropriate to their role as surrogates.

Finally, condition (4) is that the physician has made a thorough effort to communicate with the family, explaining the medical facts in a way they can comprehend, helping them understand their role as surrogate decision-makers, and enlisting the help of others in communication and mediation. Often it takes time for family members to understand the medical reality and to accept an impending death. Denial and avoidance are common methods of coping with devastating news. Healthcare professionals must of course be aware of these powerful psychological forces and offer time and assistance to the family as they adjust to the situation. Making life-or-death decisions is itself a stressful responsibility that adds to the burden the family must bear. Other professionals such as ministers and social workers are trained to deal with such situations and may provide a kind of assistance to families that physicians are not able to provide. Hospital ethics committees may offer a forum for the resolution of differences and may assuage some of the doubt and guilt that families often feel when they agree not to do everything possible. Physicians should take advantage of all these resources and should be patient and sympathetic to the needs of family, but they also have to consider the welfare of their patients. At some point they should be able to put the patient's needs ahead of the needs of the resolute family. Hospital policy should not force them to continue to inflict painful procedures on their patients when family members are insisting on it for inappropriate kinds of reasons. It is wrong on the face of it to cause one person to suffer without benefit in order to satisfy the demands or needs of another person. Physicians should be able to weigh the degree of patient suffering against the needs of the family and the importance of their involvement and, in extreme cases, protect the patient from further medical harm.

The proposal to allow physicians in some cases to act without family consent may appear to be an attack on patient autonomy and a return to medical paternalism, but it is not. Surrogate decision-making is in general justifiable as a practice that extends patient autonomy, by allowing those who should know the patient best to choose as the patient would choose. Though this indirect autonomy is an imperfect copy of the original, it should be respected when conscientiously exercised. But we must realize that unqualified acceptance of the practice may in some cases fail to promote either the preferences or the best interests of patients. When it does not work the way it is supposed to work, patients are vulnerable and may need protection. The duty not to harm must be weighed against respect for patient autonomy. The less directly autonomy is exercised and the less it is based on a patient's preferences, values, and interests, the less weight it brings to the moral scales.

If accepted, the proposal would not reduce the proper role of the family. Reasonable people may disagree about what would be best for the patient, even when applying the same criteria of suffering, quality and length of life, and so on. When this occurs, physicians should defer to the judgement of the family. It is only when the family will not deliberate in an appropriate manner – that is, try to determine either the preferences or the best interests of the patient – or when their views about suffering or the possibility of cure diverge radically from the mainstream of society (and there is no reliable evidence that the patient shared those views) – it is only then that hospital policy should allow physicians to act unilaterally to protect the patient from further harm.

III. CONSENSUS AND FUTILITY

To this point we have considered the physician's need to reach consensus with the family or other surrogate when the small benefit that a treatment provides is clearly outweighed by the pain and suffering it produces. Let us now consider those procedures that would be futile, that would provide no benefit to the patient. A treatment would obviously be futile if it would not accomplish its intended physiologic effect. Cardiopulmonary resuscitation (CPR) would be futile, for example, if it would not successfully restore a heartbeat. Even if physiologically effective, CPR would also be futile if it could not postpone the moment of death because of other untreatable conditions (e.g. multiple organ failure). Though it might produce the intended physiologic effect of restoring ventilation and circulation, the effect would not be a benefit to the patient [9].

Physicians can sometimes be reasonably certain that a procedure would be futile in this strict sense of the term. When that is the case, they should not be required by hospital policy to offer such procedures to families as an option. The justification for any medical intervention – even prescribing aspirin – is that there will be some benefit to the patient that will outweigh its burdens, that is, undesirable consequences such as pain, suffering, loss of function, and expense. Since there is no medical or ethical justification for futile treatment, there should be no institutional obligation on physicians to offer procedures that would be futile.

Generally there is no such obligation. In fact there is an obligation not to offer them, as recognized by the American Medical Association's Council on Ethical and Judicial Affairs in an opinion that physicians should not provide, prescribe, or seek compensation for services that are known to be unnecessary or worthless [1]. The only exception generally created by hospital policy is CPR, which is required for every patient suffering cardiac arrest unless a do-not-resuscitate (DNR) order has been written by the patient's physician. Since hospital policy usually requires that family members agree to such an order, the physician must seek consensus with the family before the order can be written. Thus it is possible for family members to insist upon repeated resuscitation of patients who will not benefit from the procedure and whose suffering may be prolonged or intensified.

The previous proposal was that physicians should be able in extreme situations to override the decisions of surrogates when they are based neither on patient preferences nor best interests. Though treatment should be offered so that the family can consider the patient's preferences and best interests, their decision should not necessarily be the final one. The present proposal is that, if the procedure would clearly be futile, then it need not even be offered to the family for their deliberation. If there is no medical justification for a procedure, the physician should not have to offer it to the family and make a case against it. CPR should be no different from any other treatment in this respect. Futile surgery would not be offered to a patient or family with the suggestion that it be refused. Such discussions are inherently misleading, implying that there is some benefit to the procedure – why else would it be offered? While most families will have no difficulty in the end accepting the physician's recommendations, some will interpret declining the procedure as giving up too soon, while some hope remains. They will agonize over the decision and worry that they are breaking faith with the patient. Denial of death or guilt over previous treatment of the patient may preclude understanding and acceptance of the physician's recommendations [4]. Autonomy is not enhanced by offering futile measures as options; rather it is diminished by such discussions, which are confusing and misleading and thus reduce the agent's ability to respond appropriately [11]. Whenever there is a reasonable possibility that a procedure would benefit the patient, then it should be explored with the patient or family, but not if empirical studies together with clinical evidence indicate virtually no possibility of success.

In practice it is rarely possible to conclude with complete certainty that a treatment will be futile in the strictest sense. There is almost always some chance, however small, that the procedure will work and will benefit the patient in some way. This fact leads Lantos and colleagues to the view that, out of respect for patient autonomy, physicians must always discuss the chances of success with patients or families and must achieve consensus that the probability is unacceptably low before treatment may be withheld [6]. It is of course true that absolute certainty is not possible in any clinical science, but physicians can sometimes achieve a degree of confidence that approaches certainty. It is an appropriate exercise of professional judgment to conclude that a procedure in a given circumstance is not a realistic option, that the possibility of benefit is too low to be rationally acceptable. Patient autonomy is not furthered by offering unrealistic options, especially if the family is exercising choice on behalf of the patient. Patients and families who demand treatments that are almost certainly futile do not do so because of a rational calculation of odds, but because they are denying death, expecting miracles, avoiding guilt, or expressing concern or commitment in a symbolic manner. Tomlinson and Brody are correct that offering futile treatments in such circumstances undermines autonomy by triggering and supporting these natural defense mechanisms and by suggesting – despite what one actually says – that the procedure offers hope of some benefit. A request for the procedure "will be based on misunderstanding of its benefits, and therefore will frustrate the pursuit of autonomy rather than serve it" ([10], p. 1758).

IV. FUTILITY AND HOSPITAL POLICY

Any policy that allows physicians to make decisions without family consensus on grounds of futility should be written carefully, since the term "futile" is used in different senses. For policy purposes it should be defined strictly: a procedure may be considered futile when there is no reasonable possibility that it will be successful or prolong life. Whether or not a procedure would be futile in this sense is a medical judgment that only physicians are qualified to make. When there is no reasonable possibility of extending life, the values, plans, and hopes of patients or families are not relevant to the decision. It is for this reason that the family need not in every case be consulted about futile procedures.

A treatment may also be described as futile in a broader sense: while physiologically effective, it may fail to provide a benefit that is significant to the patient. It may not prolong life sufficiently to be meaningful to the patient, or it may fail to improve an unacceptable quality of life. For example, a patient dying of a painful cancer might consider CPR to be futile if it would only prolong the terminal discomfort for a short while. On the other hand, dying patients may still have important achievable goals – the cancer patient, for example, may be expecting a visit from grandchildren or from an alienated brother or sister. It is only with knowledge of the patient's overall goals, values, and preferences that procedures can properly be judged to be futile in the broader sense (that is, that there is no reasonable chance treatment will allow any of these goals or preferences to be realized). Thus, when treatment would not be futile in the strictest sense, it should be discussed with the patient or family, so that they can weigh the chance of benefit against the additional burdens to the patient from the patient's perspective. The choice of a patient who is able to decide should be followed, even if the physician thinks the burdens would outweigh the benefits. In similar manner the choice of the family or other surrogate should be followed as long as they are trying to determine either the preferences or the best interests of the patient. Policies allowing physicians to act alone on grounds of futility must not reduce the proper role of the family.

Supposing physicians may write DNR orders without family consent, should hospital policy require that the patient or family be told that a DNR order has been written? There are a number of reasons why physicians should inform the family of their decision. It is important for all parties to understand the severity of the patient's condition and that death may be imminent, and discussion of CPR may be the best way to ensure a realistic understanding of the prognosis. It concerns a future event, not a present crisis, so that discussion may be calmer and less threatening. Moreover, many people are already familiar with the procedure from news or entertainment programming and thus may be somewhat prepared for the discussion [12]. Because of widespread familiarity with CPR, many families will expect it and be aware that it was neither discussed nor employed. Failure to inform may be perceived as a breach of communication and trust, producing resentment or even litigation [3]. To the extent that attempted resuscitation is expected, not disclosing a decision to withhold it could be deceptive. In addition, there is a symbolic dimension to CPR which may deepen any resentment at its not even being discussed [7]. Finally, the need to explain and justify will provide protection against abuse of physician discretion.

Division of Medical Humanities
University of Arkansas for Medical Sciences
Little Rock, Arkansas, U.S.A.

BIBLIOGRAPHY

1. American Medical Association: 1986, *Current Opinions of the Council on Ethical and Judicial Affairs of the American Medical Association*, Chicago.

2. Blackhall, L.J.: 1987, 'Must We Always Use CPR?', *New England Journal Medicine* **317**, 1281-1287.
3. Doukas, D.J.: 1991, Letter to the Editor, *Journal of the American Medical Association* **265**, 355.
4. Hackler, J.C. and Hiller, F.C.: 1990, 'Family Consent to Orders Not to Resuscitate', *Journal of the American Medical Association* **264**, 1281-1283.
5. Lantos, J.D., Miles, S.H., Silverstein, M.D. and Stocking, C. B.: 1987, 'Survival After Cardiopulmonary Resuscitation on Babies of Very Low Birth Weight', *New England Journal Medicine* **318**, 91-95.
6. Lantos, J.D., Singer, P.A., Walker, R.M. et al.: 1989, 'The Illusion of Futility in Clinical Practice', *American Journal of Medicine* **87**, 81-84.
7. Nolan, K.: 1987, 'In Death's Shadow: The Meaning of Withholding Resuscitation', *Hastings Center Report* **17** (5), 9-14.
8. President's Commission for the study of Ethical Problems in Medicine and Biomedical and Behavioral Research: 1982, *Making Health Care Decisions*, U.S. Government Printing Office, Washington, D.C.
9. Schneiderman, L.J., Jecker, N.S. and Jonsen, A.R.: 1990, 'Medical Futility: Its Meaning and Ethical Implications', *Annals of Internal Medicine* **112**, 949-954.
10. Tomlinson, T. and Brody, H.: 1988, Letter to the Editor, *Journal of the American Medical Association* **265**, 1758.
11. Tomlinson, T. and Brody, H.: 1990, 'Futility and the Ethics of Resuscitation', *Journal of the American Medical Association* **264**, 1276-1280.
12. Youngner, S.J.: 1990, 'Futility in Context', *Journal of the American Medical Association* **264**, 1295-1296.
13. Weiser, B.: 1991, 'A Question of Letting Go', *Washington Post*, July 14 and 15.

ANNE DONCHIN

SHAPING REPRODUCTIVE TECHNOLOGY POLICY: THE SEARCH FOR CONSENSUS

The complexities of reproductive technologies cry out for policy formulation, and in modern liberal societies the theoretically preferred path to policy is by way of consensus. But consensus has proven easier to contemplate than to emulate. Consensus in theory functions as a criterion for validating knowledge claims or norms, a device for legitimating particular sets of rules and procedures, and a strategy for justifying public coercion where rule by majority would stifle minority voices. Jürgen Habermas' vision of social consensus as a *goal* to be achieved through rational communication without compulsion [9] exemplifies one use of this principle. Rival approaches are often based on objections to this goal oriented vision of consensus. Some question the underlying assumption that societies are, in fact, made up of knowing intellects and free wills that share either a common rational discourse or what Tristram Engelhardt has termed "a robust faith in reason's capacity to discover a univocal moral account" ([7], p. 30) Some emphasize the attainability of objective truth transcending human opinion and others contend that all truths are constituted within history. But all agree that within a system of thought or action consensus can have no more than *instrumental* value. Appeals to consensus may serve to maintain or improve a system's performance, but cannot validate the system itself.[1]

Habermas faults these views because they tend to condone manipulated consensus, that is, the kind of mechanism whereby those with influence and authority appeal to consensus language to justify imposing their power over others and keeping them from recognizing their own interests. But he himself does not believe, as some do, that truths are waiting to be discovered outside of history. He shares with some instrumentalists the conviction that truths are not discoverable but only constructable from within historical discourse. Yet because he is also persuaded that human subjectivity is not wholly lost within these contextual modes of discourse, he believes that the possibility of achieving an unmanipulated consensus is not hopeless.

A related group of liberal thinkers, who specifically address consensus problems affecting public coercion, argue that within the pluralistic

H.A.M.J. ten Have and H.-M. Sass (eds.), Consensus Formation in Healthcare Ethics, 229–250

framework that marks liberal societies, efforts to achieve substantive consensus are bound to violate the interests of some groups. They advocate limiting the search for consensus to procedural agreements that exclude judgments of moral value. It is only these agreements that are appropriate for debate within the public realm. The balance of morality is left to private determination. But only a minimal procedural consensus is likely to be achievable on these terms.

Two other objections to consensus building projects have gained increasing influence. First, for many it is by no means obvious that any set of procedures can be found that does not already incorporate moral values rejected by some groups likely to be affected by the policy at issue. For the very principles that liberal discourse classifies as procedural presuppose such central values as individual liberty and eligibility to participate in the social contract; and these are morally intolerable to communitarians or utilitarian supporters of animal rights. Second, appeals to moral consensus tend to rely too heavily on idealized norms, hypothetical constructions of how prototypical people would behave under a set of speculative conditions.[2] In the thick of actual experience morally relevant conditions may obtain that could not readily be envisaged from an idealized perspective. Differences owing to the natural lottery or unjust distribution of resources are less troublesome to those who occupy the more privileged social positions. So unless public processes can be structured so that the least well off can without jeopardy speak on their own behalf, a genuinely democratic process is unattainable.

Without pressing this debate beyond theoretical distinctions to examine specific policy debates, the full implications of this controversy are not likely to become apparent. The inherently controversial activities bound up with new modes of procreation provide a good testing ground to examine consensus related issues and evaluate the effectiveness of efforts to shape public policy in a manner that achieves consensus. Debates about permissible uses of gametes and embryos outside the human body and future reproductive and genetic research involving them raise problems that are critical to a vast array of consensus seeking pursuits and bring to the fore issues about the role of law in regulating morality that are central to the development of public policy in all liberal societies.

This chapter examines the British experience in shaping reproductive policy through legislation in order to identify conditions for and constraints on consensus building pertaining to new modes of procreation. In little more than a single decade over one hundred other countries established national commissions to struggle with moral and social issues arising from new uses of reproductive technologies.[3] The British experience aptly illustrates the character of much recent debate and strategies available to those who control the dominant institutions to influence the legislative process. The procedures through which policy was forged reveals much about interconnections between morality and public policy and alliances between medical and legal institutions. Many look to Britain to draw lessons relevant to regulation in their own countries. George Annas commends the regulatory framework established in Britain as a model of quality control for the delivery of infertility services.[4] The Report of the Canadian Royal Commission on Reproductive Technologies [26] takes the British scheme for licensing clinics to be the most fitting model for Canadian legislation. But others doubt that a normative political consensus regarding the new technologies of reproduction is attainable. Engelhardt, among others, argues that the state should limit regulatory efforts to protection against fraud, unauthorized harms, and coercion ([7], p. 20). In the light of such incompatible assessments by qualified "experts" on these issues, it is worth returning to the British debates to reassess the process through which regulation was achieved. Given the highly controversial issues at stake, how could the legislative route devise a regulatory scheme that wins public support? What social aims are apt to be advanced, shunted aside, neglected? Can a fair regulatory framework be imposed in the absence of consensus that adequately protects the health of those who use fertility services and the children they will bear? Is protection of both the well being and self-determination of the users of these services compatible with the pursuit of other aims, most notably the protection of embryos? What policies should people support who are committed to fostering a society that fully respects the interests of all affected including those who lack voice in shaping policy?

Attending to only a subset of these questions, I will argue that the leading contenders in the British controversy brought to the embryology debates preexisting agendas that they were not prepared to modify more than was absolutely essential to achieve legislative closure. The leadership of the principal factions in the embryology debate was closed to unanticipated possibilities and new points of view that consensus seeking requires. The leading political players already had a settled view of their vital interests and actively manipulated the parliamentary processes to advance them.[5] The issue of consensus was reduced to a tool by the principal parties to the debate to press their specific agendas, by those opposed to the Warnock recommendations to promote their own cause, and by the Government to suppress the interests of groups lacking substantial political influence. My interpretation of events suggests that the liberal state's purpose in pushing a consensus model may aim primarily to legitimate the authority of the "upper culture" and suppress the norms and values of subcultures.[6] My concluding remarks will accordingly focus on the need for alternatives to consensus, particularly a more pragmatic model of agreement that is fully responsive to interests beyond those of individuals and the state.

THE BRITISH REGULATORY EXPERIENCE

Britain was both the first country to achieve a human birth through in vitro fertilization (IVF) (1978), and the first to initiate systematic regulation of embryo research and fertility services. In 1982 the British government established a committee of inquiry "to examine the social, ethical and legal implications of recent, and potential developments in the field of human assisted reproduction." ([33], vi). Two years later the Warnock Committee issued its legislative recommendations. Many considered the report ill-timed, either it was too early or already too late.[7] But they agreed that consensus was what mattered. M.P. Leo Abse [1] maintained: consensus would have been more easily achieved had the committee convened four years earlier (before the ascendancy of the religious right) and invited public comment by issuing an initial interim report. Philosopher Richard Hare [10] argued: in the absence of a public consensus, regulatory recommendations would only enshrine inconsistencies embedded in public sentiment. So better to wait until a set of socially approved recommendations emerged that would be surer to promote social utility, to him the key ethical consideration. Mary Warnock, Chair of the committee that drafted the report, also a philosopher, countered Hare's utilitarianism by arguing that legislators could rely on the public's felt responses and intuitive revulsion to specific technological innovations. The principled consistency utilitarians seek only promotes "insensitivity to the kind of inhibitions and scruples which are at the center of morality" ([32], pp. 247-8). Her supporters called attention to the most "disgusting" case scenarios: implantation of human embryos in animals and the creation of chimeras. They claimed that swift legislation was vital to halt such practices.

Neither Hare nor Warnock, however, fully appreciated either the multiple sites of power at the source of conflict or the irreconcilability of the moral ideologies at work within the controversy – one supported by the apparatus of Government, the other mounted by the religious right. Nor did they recognize the presence of other legitimate interests that, lacking access and voice, were unlikely to be heard.

In the end Abse, Hare, and Warnock were all partially vindicated. By the time the Committee had completed its deliberations Warnock had come to consider the notion of a consensus morality an untenable myth ([33], xi). And contrary to the expectations of Committee supporters, another five years passed before the Government was willing to risk bringing its Embryology Bill to Parliament (1989). Like earlier British efforts to relate controversial legislation to morality, notably The Wolfenden Report on homosexuality [35] and the Williams Report on pornography [34], consultation with public opinion was expected to reduce controversy and bring consensus. However, the Report's particular mix of utilitarian concerns and deference to popular feeling did little to dampen political controversy. The fact that the Warnock Committee pursued a different course than either the Wolfenden and Williams Reports is relevant too. Both of the earlier commissions had bypassed the search for substantive consensus and followed a procedural route, distinguishing between the public realm where morality is enforced by law and a private sphere within which individuals are left free to act on their own moral preferences. The Warnock Committee chose to incorporate a fuller set of substantive values.[8] Debate on the Embryology Bill demonstrated to all who were not already persuaded that the anticipated consensus had failed to materialize. The parliamentary process opened up deep cleavages in British opinion. The Warnock Commission, itself, contributed to its polarization.

The way the Committee framed and ordered the discourse, both morally and scientifically, intensified divisions by structuring debate around certain narrowly defined issues. Future practices were specified in a manner that enshrined the state of scientific development prevailing at the time of their deliberations. During the six year delay that was to consolidate a consensus, conflict intensified and stiffened around antagonistic moral ideologies. Issues relating to embryo research were given undue prominence to the neglect of other concerns that had grown in importance since the Warnock Report initially appeared. During debate, older unresolved moral controversies resurfaced (abortion and donor insemination, particularly), and these conflicts became entangled with new problems calling for very different remedies.

In the interim between the release of the Warnock Report and the Government's Bill the leading medical organizations interested in pursuing embryo research grasped the initiative. The Medical Research Council and the Royal College of Obstetricians and Gynecologists, established a Voluntary Licensing Authority (VLA) incorporating their own guidelines and implementing informally certain recommendations of the Warnock Committee, namely the licensing of clinics engaged in embryo research and practice and the collection of data on embryo research and laboratory fertilization. That the VLA took on these particular functions is attributable to several factors including the relation between private sector fertility services and public sector healthcare. Most obvious, of course, was the medical profession's perceived need to protect a controversial area of scientific activity, both by displaying a semblance of control [36] and by establishing its own title as precursor to the statutory licensing body to be established by legislation. Both these factors are interrelated.

Only two of the thirty-eight IVF centers approved under the voluntary scheme received direct funding from the National Health Service (NHS).[9] Privately funded facilities were not subject to the oversight procedures governing public ones. Hence the emphasis on licensing as a method for gaining access to private facilities to collect data and integrate them in a common informational network. Also, fertility clinics were being faulted because new clinical practices in the private sector were impacting economically on the public sector – for example, by increasing the number of low birthweight infants requiring care in NHS facilities. In 1987 the VLA brought under its scrutiny newly emerging practices that were attracting increased public attention since they involved significant medical risk and generated moral and social controversy: the number of ova or embryos transferred in a single procedure, the donation of ova by known donors and selective reduction of multiple fetuses (called by some opponents "feticide"). Though the VLA had no legal authority or powers of enforcement beyond the withdrawal of a clinic's license, many presumed that peer pressure would suffice to insure clinic compliance with its guidelines. But the allure of commercial success proved more powerful than the disapproval of peers. The VLA's quasi-legitimacy was challenged by clinics that resented this intrusion into their "clinical judgment." Some practitioners[10] sought to divert public discussion from social and political issues requiring collective management to the issue of clinical freedom to manage individual cases without outside interference. These practitioners argued that restriction on the number of embryos to be transferred reduced the chances open to individual women to achieve pregnancy. However, widespread attention to the economic rewards sown by some clinicians led others to wonder whose interests were being served under the banner of unrestricted freedom of clinical judgment.[11]

Nonetheless, the activities of the Licensing Authority were widely praised when the Embryology Bill finally made its parliamentary debut. Even the most vehement opponents of embryo research rarely criticized its regulatory efforts.[12] So it was able to play a significant role in structuring parliamentary debate and shaping practices to be regulated under the Act. The shape it had already assumed by embodying a particular interpretation of the regulatory provisions laid out in the Warnock Report would not be undone by legislation. On the contrary, its organizational structures and procedures provided a model to integrate into a body with expanded enforcement powers. Through this strategy the medical profession was able to institutionalize IVF technology well before the establishment of formal institutional procedures. The VLA also influenced the practice and expectations of researchers, and these influences would also endure.[13] On the eve of legislation the VLA renamed itself the Interim Licensing Authority (ILA), a move which further secured its identity as *the* regulative body described in both the Warnock Report and the Embryology Bill. Since all of the members of the ILA, both medical and "lay," were selected by the two sponsoring medical bodies, their continuing influence is assured [27].

The intensely adversarial character of the debate had already been established long before the ILA organized its lobbying campaign. While the government waited for consensus to emerge, opponents of the Warnock recommendations seized the initiative. Within a year after their Report, Enoch Powell, M.P., introduced into Parliament the Infant Life (Protection) Act (1985), linking the issue of embryo research to abortion controversy. Unlike earlier objections to IVF that had focused on fears that laboratory manipulation might cause developmental abnormalities in embryos, the backers of Powell's Bill claimed that embryo research violated the *moral* status of the early embryo irrespective of its subsequent development. His Bill sought to ban all embryo research, permit IVF only after securing the written permission of the Secretary of State, and require fertilization and transfer to the woman of all ova recovered (despite risk to her and the fetuses). The large body of support he was able to marshall for his Bill demonstrated the effectiveness of a new political strategy which broadened the base of the "new right," a politically unaligned moral lobby composed of traditionalists and neo-conservative groups [1,36]. Anti-abortion rhetoric was extended to embryo research and political support was mobilized around appeals to the sanctity of human life from conception. By so linking two issues that posed distinctive moral and social problems, the new right restructured the agenda for all subsequent debate, and caused the status of the embryo to be elevated to prime position. Indeed, the Warnock Committee itself had invited this outcome by identifying "the value of human life" as *the* most significant of all the issues it considered ([33], p.xvi). But the Powell proposal forged the crucial link connecting this hierarchy of moral priorities to the already explosive issue of abortion. Public debate came increasingly to focus around what one commentator has called "an over-individualized notion of the embryo" that ignores the site of its origins and the necessary conditions for its subsequent nurture and development [36]. Two distinctive uses of "human," the descriptive and evaluative, were thus conflated, dissolving the distinction between biological and social meanings of embryonic life.

Recognizing that a ban on embryo research would severely hamper IVF practice and development, researchers and medical practitioners belatedly assembled their own lobby (PROGRESS) to fight the Powell Bill. But to the consternation of the Government and other Warnock Committee supporters, Powell's Bill won majority support in the House of Commons. It failed to become law only because of time limitations governing private members' bills.[14] Fortified by the Bill's near success, organizations that opposed the Warnock proposals and backed Powell – the Catholic Church, the Society for the Protection of Unborn Children (SPUC), and LIFE – continued to wage a militant, tightly organized and well funded campaign to halt embryo research, to limit possession of an embryo to use for a specific woman and to recriminalize many abortions (since 1967 abortion up to twenty eight weeks has been available with the consent of two physicians). So when the Government finally decided to bring its own bill to Parliament opponents were ready and waiting. The Government, however, also contributed to the polemical character of the ensuing debate.

The Government's Bill was a lengthy and detailed document incorporating many potentially controversial provisions which merited consideration and approval on their own terms. But as introduced it discouraged separate examination of such provisions and precluded their selective adoption. To accommodate the anti-abortion/anti-research lobby, it included two alternate and mutually exclusive clauses, one permitting embryo research for up to fourteen days and the other prohibiting any research on embryos unless they were subsequently transferred to the woman from whom the ova were taken. Members were allowed a "free vote" only on this clause. (Free votes are usually only given on matters of conscience. Otherwise MPs are expected to adhere to party policy.) Any other changes in the document required a cumbersome amendment process.

The Government further veered from neutrality by allowing only abortion-related amendments to the Bill.[15] A move to detach abortion debate from the Bill succeeded in the House of Lords, only to be restored by the Government when the Bill came to the Commons – where members were also given a free vote on abortion. This move effectively diverted attention from all other concerns in the original Bill with the sole exception of embryo research.

Emboldened by their successes, opponents of the Bill laced their rhetoric with misrepresentation of scientific findings, personal abuse of opponents, biblical allusions, and claims to divine authority. The Archbishop of York, a supporter of embryo research, was labeled publicly as a "disgrace to Christianity," "committed to a pro-death cause" [11]. Opponents accused the embryo research group of "nazism" – "(they) want to have embryos in order to detect chromosomal and genetic disorders, not primarily in order to cure them ... but to be able to detect these defects and to kill them" [5]. They focused so narrowly on the embryo that the woman who provides the essential conditions for its development was reduced to invisibility. By such devises they polarized and oversimplified virtually all issues tied to the Warnock recommendations.

Research advocates, for their part, abetted polarization by focusing their own critique of the anti-research position around the status of the early embryo and exploiting their scientific prestige for political advantage. Their use of moral polarities (abortion involves a conflict of evils/embryo research involves using the conceptus for the possible benefit of others) also tended to disregard the intimate connection between pregnant women and fetal life. Their analysis of embryonic development and some dimensions of moral relationships might under other circumstances have injected fresh perspectives, but in the context of the parliamentary debate it only intensified dichotomous thinking.

The status of the embryo as an autonomous entity having become the debate's central focus, those who favored abortion rights but opposed embryo research could find no footing. Debate was so polarized that they were no longer perceived to be offering an alternative *moral* position but were summarily dismissed as *logically* inconsistent. Moreover, since many embryo research supporters were also among the strongest supporters of women's abortion rights, abortion defenders found opposition to embryo research too politically risky. PROGRESS, the pro-research lobby which made its first appearance with the Powell Bill, was hastily reconstituted to lobby Parliament and respond to the barrage of literature opponents were heaping on MPs.[16] Their own literature replicates the dichotomous discourse of their opposition – for example: "While the moral arguments, of their nature, remain subjective, the scientific arguments for continued research are cogent and objective" ([24], p. 1). By controlling scientific language used by medical authorities (the term "pre-embryo," for instance) they were able to influence both members of Parliament and the educated public.[17] Under the heat of debate few at the time could see such maneuvers as "linguistic engineering" in the service of genetic engineering as retrospective commentary put it ([20], p. 75).

In several respects, institutionalization of the ILA provided a vehicle for the embryo researchers dominating it to control the discourse about regulation. They adroitly undercut objections to regulatory proposals by citing their own procedures as evidence of adequate oversight. For instance, after opponents of embryo experimentation criticized researchers for using human embryos where animal embryos could have been used instead, the ILA expanded its guidelines for approval of projects to include a statement explaining why a particular experiment could *not* be carried out on animal embryos. Also, when opponents pointed out the low success rates of IVF, research supporters changed the description of their patient population from "desperate infertile women" to "parents burdened with the care of seriously ill children" and emphasized the need for further research on human embryos to improve the IVF success rate following pre-implantation genetic screening, so fewer children would be born with sex-linked genetic diseases. The Warnock Report had characterized the aim of embryo research as the alleviation of infertility but, as debate proceeded, researchers were able to shift the emphasis to the control of genetic anomalies. Preimplantation diagnosis, a highly charged issue in France and Germany, was in Britain turned on its head and employed as a principal argument *favoring* IVF and embryo research. In response to fears about genetic engineering, embryo research supporters in Parliament emphasized differences between alteration of genetic structure by inserting replacement genes into germ cells (which they did not contemplate) to genetic screening, somatic cell therapy, and preimplantation genetic selection (which they did). To allay any remaining anxieties they linked the manipulations they supported to natural (less efficient) processes of embryo wastage and natural selection. Different moral concerns in British, French, and German debate can be explained in part by differences in cultural experiences.[18] But the prestige of the Warnock Committee, establishment of an interim licensing body by the leading medical organizations, and the widespread perception that the Licensing Authority was its creation are also factors in the British outcome.

The medical establishment also managed to block amendments to the Bill that would have expanded the scope of regulation. Those favoring amendments pointed out that the Bill did not reflect the current state of scientific practice, but conditions prevailing at the time of the committee's deliberations. Subsequently, clinicians had developed and applied a number of innovations, most notably gamete intra-fallopian transfer (GIFT), now employed as frequently as IVF and regulated in other countries. As the Bill was framed, GIFT, which does not involve laboratory fertilization, falls under regulation only if donor gametes are employed or if the hamster egg test is used to assess the fertilizing capacity of the sperm.[19] In response to controversy about the exclusion of GIFT, sections were added to the Bill to allow for possible extension of coverage in the future. Since implementation of the Act a number of clinicians have complained that the exclusion of GIFT perpetuates an inferior standard of care while exposing women to the same risks as IVF. But no action has been taken to bring GIFT under scrutiny of the Authority, possibly because these objections do not address the main thrust of the Act, the protection of embryos [16]. The exclusion of GIFT from the regulatory framework illustrates the extent to which both the Government and the medical establishment were blinded to serious ethical issues pertaining to the women whose bodies are put at risk. For like IVF, GIFT is generally accompanied by the use of superovulatory drugs to stimulate the production of multiple ova (risking ovarian hyperstimulation syndrome and possible long-term effects). Both involve risks associated with multiple pregnancy; both carry a significantly higher than normal rate of infant morbidity and mortality; and both involve considerable financial and social costs (treatment fees are comparably high, success rates low, and the expense of caring for impaired newborns substantial). Moreover, the incidence of tubal pregnancy and the risk of multiple pregnancy are even higher with GIFT than with IVF. And since surplus eggs must be discarded the retrieval process must be repeated for each subsequent cycle. By excluding GIFT, practitioners can easily circumvent regulation, even without medical justification, by opting for GIFT over IVF.

Many participating in the parliamentary debate bemoaned the lack of available options and some vowed to fight back when the Bill went to committee, but it emerged with few changes. Feminists in Parliament were placed in an awkward position since the issues of embryo research and abortion had been joined so crudely. On the free vote, however, abortion opponents were defeated and the clause prohibiting embryo research lost by a substantial margin. So with few exceptions the Bill passed without significant revision. The most noteworthy change resulted from the attachment of abortion amendments. Despite a public perception that the abortion law had been tightened by the reduction in the time limit from twenty eight to twenty four weeks, the legislation actually had the opposite effect since exceptions to this time limit were built into the Bill. The anti-abortion lobby conceded that their cause had suffered a disastrous setback [31]. In 1991 the ILA became the Human Fertilisation and Embryology Authority (HFEA), the statutory body governing access to gametes and embryo research. Its scope is somewhat greater than the ILA's had been since it controls the storage of gametes and embryos and all forms of donor insemination, inspects the centers, and maintains a central register of information. But the framework within which it is to operate was little changed.

CONSENSUS BUILDING: WHO COUNTS?

The predominantly negative reaction of most feminists to the Warnock proposals and the subsequent debate regarding them helps expose the predominant moral emphasis that guided many of the Committee's recommendations. Disproportionate weight was given to people's psychological feelings about reproductive innovations with little recognition of the extent to which such feelings are an artifact of the authoritative voices of governmental committees and their particular ideological construction of family life.[20] A narrow focus on some implications of new reproductive practices neglected others. The exclusion of GIFT from regulatory scrutiny was not the only provision of the Act that exhibited a lack of regard for the women who use these treatment services. Though the Warnock Committee had called for national guidelines on the organization of medical services, the Act made no explicit provision for them. It failed to address even the most blatant inequities in the provision of fertility services which are concentrated in the most affluent geographical areas and cater to patients with substantial private funds.[21]

Debate about laboratory insemination focused far more on concerns about the need to protect the blood line than on the interests of women who receive sperm from anonymous donors and the children they bear. Following debate on artificial insemination in the House of Lords, Mary Warnock, having recently been granted a life peerage, came out in favor of using sex selection techniques to assure a male heir to those with hereditary titles [28]. The patriarchal interests of Scottish clansmen did not fare so well. Their demands for disclosure of the biological father's identity (to insure that clan membership followed the biological line) were often greeted with scorn and derision. Though a few favored disclosure for other reasons, the interests of those wishing to preserve the fiction of the "natural family" prevailed. When the Government's Bill emerged from the Lords committee only one substantive change had been made: to allow IVF babies born to the aristocracy the right to inherit titles and take up seats in the House of Lords. In this respect the legislation reflects a distinctively British cultural preoccupation, a peculiar convergence of pre-capitalist and capitalist forms of patriarchy, each embodying distinctive conceptions of what is to count as a family.[22] The debates made such sexist agendas explicit, but they were implicit already in the Committee's report.

Rosalind Petchesky in the foreword to the British edition of her book, *Abortion and Women's Choice*, captures succinctly the sense of outrage that shaped the response of many feminists to the Report and subsequent legislation.

The Warnock report may be read as an ideological document that lays down the terms of a neo-liberal, utilitarian "fertility contract." In its very title, *A Question of Life*, the Report overdetermines the issue of whether or not or how human beings ought to "bring life into the world" or "destroy life" as the central moral issues in fertility control debates. Other ways of framing what constitute 'the moral issues' – for example, the distribution of responsibility for children after birth, the distribution of power over reproductive decisions, the conditions of informed consent, the health needs of women – are nowhere considered. In particular, the Warnock Committee immediately reduces the "morally relevant" question in reproductive decisionmaking to how to regard embryos or "the value of human life." ([22], pp. xv-xvi)

The parliamentary process offered little opportunity to reframe either the issues or the terms of the "fertility contract" laid down by the Warnock Report. Questions of paternity and physician responsibility in manipulating gametes and embryos so dominated debate that little attention could be diverted to women undergoing treatment who tended to be conflated with "the infertile couple," a rhetorical twist that has become common usage in medical and legal circles, effectively erasing the agency of the women subject to medical intervention [17,27]. Women were given no choice but to accept a fertility contract that gives to some limited access to new modes of reproduction. The lopsided favoring of embryonic life over the lives of women patients, the manifest hostility to unmarried women who seek fertility services, and the exclusion of GIFT from the scope of the Act together demonstrate a willful disregard of patient interests. The patriarchal bias of the report and parliamentary debates was even more conspicuous where they addressed access of unmarried women to fertility services. This issue aroused nearly as much venom and animosity as embryo research and abortion. By way of compromise, a vaguely worded provision, commonly referred to as the "welfare of the child" clause, was included that requires treatment centers to take into account "the child's need for a father" and to balance the interests of prospective parents and the potential child on a case by case basis.[23]

In the wake of legislation the HFEA has struggled to fill obvious gaps in the law and has sought to include in its membership a broader spectrum of the public. The Code of Practice governing the implementation of the Act provides a much more flexible guide to practice that is more readily subject to revision in the light of changing experience. The HFEA has conducted numerous surveys and solicited the opinions of diverse groups on issues such as sex selection and donation of ovarian tissue.[24] Their adaptability in interpreting and applying the law, their accessibility to the public, and their openness in communicating information about their activities have all contributed significantly to the law's acceptability and the international interest it has attracted. British patients are likely to have access to more detailed and accurate information and higher, more uniform quality treatment procedures than is available, say, under the free market system that still prevails in the U.S.

THE AFTERMATH: SOME DOUBTS ABOUT CONSENSUS MODELS

Though some characteristics of the political process that led to regulation of fertility services in Britain are peculiar to British culture, most are generalizable. In particular, the three distinct interests in evidence during the British legislative debate – researchers, an anti-research/anti-abortion lobby,[25] and government – are likely to surface virtually anywhere, though their relative strengths will vary. Some claimed that the British Government had little interest in *how* matters were resolved as long as closure was achieved, but the extent to which the Government orchestrated the legislative process belies this assessment. Arguably, any government has an interest in maintaining stable patterns of family arrangements (though constructions of the family will vary), at least to prevent the burden of child care and maintenance from falling preponderantly on state agencies, or to prevent the judicial system from becoming overburdened by competing claims to parental rights. No democratic state is likely to push for comprehensive legislation if it can achieve its ends in other less cumbersome ways. It is more apt to legislate in some areas (where the threat to preferred family norms is more immediate, like commercial surrogacy) than in others (where procreative innovations can be more easily adapted to approved norms: DI, IVF, GIFT).

The interests of physicians and researchers who deal with infertility are obvious, since their professional futures are tied to rapidly developing medical specialties that exploit scientific advances in embryology and molecular genetics. They are inclined to ward off legislation which they see as intruding on their own domain. When the threat of legislation increases, they seek out ways to demonstrate their capacity for voluntary "self-regulation" by devising "codes of conduct" and, if pushed further, oversight bodies consisting largely of their own membership, in order to retain as much control as possible over their own practices.

In some other countries the anti-research lobby (in Britain made up largely of Roman Catholics and Christian fundamentalists) is not so likely to be the only vocal group with reservations about recent developments in reproductive medicine, genetics, and embryology. But others who question the commitment to high-technology infertility intervention seldom have an effective lobby. They are rarely well organized, unlikely to be represented on commissions since officials seldom recognize them as "key" interest groups,[26] and rarely have sufficient funding to mount a forceful campaign to resist either the medical lobby or the fundamentalist opposition. Feminist groups, environmentalists, public health professionals, racial and ethnic minorities, and handicapped people have all expressed serious doubts about the benefits to be derived from high-technology infertility intervention. Their influence varies from country to country and each group is apt to view legislation differently. But as the institutionalization of reproductive practices shifts the criteria for procreative decisionmaking from affected individuals to the institutions that determine who will have access to the techniques and on what terms they will be provided, groups without political influence or representation are likely to be further undermined. The particular mode of institutionalization adopted in individual countries will determine who has access to medically assisted procreation, how gametes and/or embryos will be selected, and to whom medical practitioners will be responsible. Institutionalization has far more diverse effects than the specific regulatory measures adopted, influencing the structures of public discourse surrounding reproduction and community acceptability of changing procreative practices.

So the process and outcome of comprehensive regulation raise profound questions for several of the groups that have stakes in the process, particularly insofar as governmental authorities who have the power to guide the process appeal to a consensus model to achieve their goals. Pressures to force a *national* consensus under government auspices are likely to overlook the interests of groups lacking substantial political influence. Women's interests, particularly, may not fare better within institutional structures under government control than under market control.

The push for consensus among *international bodies* raises still different doubts about the drive toward consensus. Jonathan Glover's report to the European Commission [8] calls into question the search for a uniform European policy. The report notes the extent to which different countries have very different religious and ethical traditions that are not likely to yield readily to pressures to harmonize. It is better, the report claims, to allow a European moral outlook to evolve, if at all, out of a slow process of mutual discussion and focus on the reasons national regulatory bodies give for their decisions rather than on their national character. Though Glover sees little prospect of reaching international consensus at the level of moral principles, the report holds out the promise of eventual agreement by aiming at a reflective equilibrium between principle and practical intuitions.

Pan-European consensus, even on the terms of those who control national legislation now seems very remote, since institutionalization is proceeding so diversely in different countries and will have far more pervasive effects than specific regulatory measures alone. For it also influences both the structures of discourse about the meanings of reproduction and the acceptability of changing procreative practices. As we have seen, the institutional structure into which British legislation was fitted had already been shaped by a particular configuration of interests. As other countries move toward institutionalization reproductive techniques are circumscribed within still different structures, and these will largely determine regulatory outcomes.[27]

The distinctive ways each country shifts the criteria for procreative decisionmaking from the individuals who are the subjects of technological interventions to the institutions which determine access to the techniques intensify these differences. From the perspectives of groups that are not well served by such institutions, the discourse of consensus may seem but a pretense to consolidate power and control in the hands of the elites who dominate the institutions legitimated by legislation and a cynical maneuver to justify measures disciplining those who are marked as deviants by these policies. Engelhardt is right about the impossibility of attaining consensus where there is widespread doctrinal disagreement. But he is wrong, I am convinced, in assuming that the only alternative is minimal governmental regulation. For there are other interests at stake that extend beyond the preferences of the prospective parents who are fortunate enough to gain access to fertility clinics. There are social goods to be fostered that are not reducible to the interests of the individuals directly involved. Power imbalances between patient and physician and the interests of third parties and potential children also need to be taken into account. For practitioners control access to the resources that permit conception, determine the boundaries of permissible risk, and make judgments extending well beyond both their own technical expertise and the consent granting domain of their patients.[28] So even in the absence of consensus, government has a responsible role to play even if it cannot do so under the mantle of consensus. What justice minimally requires is not the achievement of consensus, but a genuinely public discourse in a forum open to all, that includes in its ranks not only experts but also those whose knowledge stems from experience and reflects the standpoints of marginalized groups. Such a discourse will take into account differing appraisals of the burdens and benefits that the new technologies bring in their wake and the distinctive norms and values of all groups that have a stake in reproductive practices, including representatives not only of the dominant culture but of subcultures as well, so that as practices are transformed the voices of the disenfranchised will be heard.

Department of Philosophy
Indiana University
Indianapolis, Indiana, U.S.A.

NOTES

1 I am indebted here to Lyotard's formulation ([18], p. 60ff) and especially to his interpretation of the ideas of Niklas Luhmann who has also influenced Habermas (and v.v.). On the issue of consensus Habermas argues against both Luhmann and Hannah Arendt.

2 Both of these objections are explored in greater detail by Kuhse [15].

3 Ireland, Denmark, Germany, Portugal, Norway and the Australian states of Victoria and South Wales banned embryo research either by legislation or by imposition of a moratorium. Spain passed legislation permitting it but regulated treatment services more extensively than Britain. For further detail see the table of international comparisons in Morgan and Lee ([20], pp. 86-87) and the comparative analysis of over 100 reports of special commissions that considered regulation of the technologies in Knoppers and LeBris [14].

4 See also Wagner and Stephenson, ([30], p. 20). Appeal to quality control may harbor a number of ambiguities. As the HFEA functions two aspects of quality control are paramount: accessibility to services and record keeping. Access is limited by both physical and social criteria including evidence of infertility and genetic impairments and suitability which takes into account maternal age and marital status. The records maintained include treatment risks and success rates of various techniques and the identity of donors which is zealously guarded.

5 For an insightful discussion of the distinction between consensus and compromise see Moreno ([19], p. 45 ff.).

6 Bianchi [4] makes a strong case for this position. His view is particularly interesting in connection with appeals to the notion of a social contract to support the primacy of consensus as a devise for reaching agreement that overcomes the partiality of majoritarian rule. He accepts the characterization of the social contract as a legitimating device but then seeks to show that it imposes on the underclass a tyranny comparable in its force to the tyranny of the majority. Helga Kuhse [15] calls attention to other shortcomings of the consensus model that intersect Bianchi's at some junctures, though she points principally to exclusion from the social contract of future people (including embryos) and nonhuman animals.

[7] For over a decade before establishment of the Warnock Commission articles had been appearing in the scientific journals raising ethical and social questions about the new techniques. Initially, moral objections to IVF focused around fears that laboratory manipulation of embryos might lead to serious developmental abnormalities and the potential for innovative forms of surrogacy. IVF births subsequently showed the first fear to be unfounded. The second was the subject of legislation in 1985.

[8] The distinction between procedural and substantive consensus is not as clearcut as some supporters of the distinction such as Engelhardt [7] suggest. For a particularly balanced discussion of the distinction see Moreno ([19], p. 41 ff).

[9] In two clinics within the National Health Service that provide treatment the waiting lists have been so long that a woman would be more likely to become pregnant while on the list than under treatment. Unless explicitly excluded, however, diagnostic procedures and the cost of drugs (which may add up to as much as $1600) may be covered by NHS though coverage varies from locality to locality.

[10] Ian Craft, director of Humana's fertility unit, was particularly vocal in this debate which brought to light the fact that some practitioners were reported to be making over £500,000 a year off their infertility practice.

[11] Frances Price [23] discusses these issues more fully. She also includes references to articles by a number of participants in this debate. My own discussion here owes much to hers.

[12] However, the Licensing Authority turned down only one research proposal during its first four years of existence. In 1989 alone 53 projects were undertaken. See their publication *IVF Research in the UK*, (1989).

[13] The Act provided that at least half the members of the Human Fertilisation and Embryology Authority (HFEA) be non-physicians who have no direct concern with funding embryo research. Still, six of the twenty-one original members of the HFEA served on the ILA and nine former ILA members continue to serve as HFEA inspectors of clinic facilities.

[14] A private member bill is one introduced into Parliament without the endorsement of the majority party.

[15] In the "long title" of the Bill these words were incorporated: "to make provisions in connection with human embryos and any subsequent development of such embryos." Though this title is roomy enough to accommodate virtually any concern about human life, only abortion amendments were actually permitted under it.

[16] For example, [2]. Researchers had obviously been working diligently behind the scenes in the interim, conducting a well organized, low key campaign to conscript MP support. A third of those who spoke out in favor of embryo research during the 1989-90 debates mentioned direct contact with the scientific community. Only 1 in 12 of those opposed made any reference to such contact. My appreciation to Michael Mulkay for pointing this out ("Changing Minds about Embryo Research," unpublished manuscript).

[17] The term, pre-embryo, was used in the Warnock Report itself. It refers to a period in the development of the embryo up to about fourteen days after fertilization. The fetal cells are not yet differentiated from the cells that will constitute the placenta and other supporting tissue, and the primitive streak has not yet appeared. The Report recommended that research should be permitted up to that point only. Beyond it the embryo was to count as a "potential human being." Some within the scientific community have argued that, though some distinction may need to be made for legal purposes, the point at which the distinction is made is arbitrary [20]. But by adopting the term, the early embryo is effectively transformed into a non-embryo, thereby diverting both legal and moral objections to embryo manipulation.

[18] Though some anti-research MPs raised the specter of nazism in their speeches they were seldom taken seriously by their opponents, perhaps because they tended to extend the accusation to all intervention even remotely suggesting a preference for a child free of genetic disease. The anxiety about nazi-like eugenic practices that lingers in France and Germany and is refueled by the prospect of state sanctioned genetic intervention lies outside the consciousness of the British.

[19] Use of the hamster egg test falls under scrutiny of the HFEA since the procedure counts technically as cross-species fertilization. At the time of a 1993 survey only one center has a license for performing hamster penetration tests [16]. Several other potentially controversial procedures which do not involve laboratory fertilization also fall outside the scrutiny of the Authority including research on ovarian cells and the techniques of embryo splitting.

[20] This was noted in a response to the Warnock Report by the Oxford University Women's Studies Committee, 1984 (unpublished).

[21] Shortly before the British legislation went to Parliament the private health care chain, Humana, opened a for-profit facility in London equipped to accommodate two thousand (private paying) patients a year.

[22] I owe this insight to Hilary Rose ([25], p. 184).

[23] Despite the ostensible intent to protect children's interests, Parliament gave scant attention to their well being after birth. Little notice was taken of studies indicating that children who grow up in lesbian families are not distinguishable on psychological tests from children reared in heterosexual families. Nor was any importance attached to adult children's interests in tracing the identity of their biological fathers, though this option is generally available to adopted children. Far greater weight was assigned to the desires of social parents to shroud the practice in secrecy and preserve the anonymity of sperm donors. Though it was recognized that information identifying donors should be retained by the HFEA for possible use in subsequent genetic screening, safeguards preventing unauthorized disclosure were drafted initially in such restrictive language that access was barred even to professionals designated to perform screening by the legal parents! In 1992 subsequent legislation was passed modifying the confidentiality provisions of the original law.

[24] See, for instance, the annual reports issued by the HFEA.

[25] Hilary Rose reports [25] that one country, Norway, has managed to avoid entanglement of embryo research regulation with abortion controversy.

[26] Rose offers some insightful comments on the composition of such commissions, in Britain, particularly ([25], p. 180). Wagner [29] and Wagner and Stephenson [30] argue forcefully for inclusion of a public health perspective in policy regulating NRTs.

[27] France passed legislation in 1994 after a long delay. However, the medical establishment's voluntary scheme for managing the collection and distribution of sperm had already been in place for a number of years. Its design and implementation vividly illustrate the interplay between technology and cultural practices.

[28] Simone Novaes cites some fascinating examples including the role assumed by practitioners in the French CECOS Federation in preventing the deliberate transmission of genetic disease ([21], p. 215).

BIBLIOGRAPHY

1. Abse, L.: 1986, 'The Politics of In Vitro Fertilisation in Britain,' in *In Vitro Fertilisation Past Present Future*. S. Fishel and E.M. Symonds, (eds.), IRL Press, Oxford, pp. 207-213.
2. All-Party Parliamentary Pro-Life Group: 1989, 'Upholding Human Dignity: Ethical Alternatives to Embryo Research', London, U.K.
3. Annas, G.J.: 1994, 'Regulatory Models for Human Embryo Cloning: The Free Market, Professional Guidelines, and Government Restrictions,' *Kennedy Institute of Ethics Journal*, **4**, 3, pp. 235-249.
4. Bianchi, H.: 1994, *Justice as Sanctuary*. Indiana University Press, Bloomington.
5. Cherfas, J.: 1989, *Science*, **246**, 12/12/89.
6. Engelhardt, H.T., Jr.: 1986, *The Foundations of Bioethics*, Oxford University Press, New York and Oxford, U.K.
7. Engelhardt, H.T., Jr.: 1994, 'Consensus: How Much Can We Hope For?,' in *The Concept of Moral Consensus*, K. Bayertz, (ed.), Kluwer Academic Publishers, Dordrecht, Netherlands, pp. 19-40.
8. Glover, J. *et al*: 1989, *Fertility and the Family: The Glover Report to the European Commission*, Fourth Estate, London.
9. Habermas, J.: 1976, 'Hannah Arendt: On the Concept of Power', in *Philosophical-Political Profiles*, Heinemann, London, U.K., (1983), pp. 171-187.
10. Hare, R.M.: 1987, '*In Vitro* Fertilisation and the Warnock Report', in R. Chadwick, (ed.), *Ethics, Reproduction and Genetic Control*, Croom Helm, London.
11. House of Lords Official Report: 1989, **513**, (11, 12/7/89), London.
12. *Human Fertilisation and Embryology Act 1990*, Chapter 37, HMSO, London.
13. Interim Licensing Authority (ILA) for Human in vitro Fertilisation and Embryology: 1989, 'IVF Research in the UK, 1985-1989,' London, U.K.
14. Knoppers, B.M. and S. LeBris: 1991, 'Recent Advances in Medically Assisted Conception: Legal, Ethical and Social Issues,' *American Journal of Law and Medicine*, **17**,4, pp. 329-361.
15. Kuhse, H.: 1994, 'New Reproductive Technologies: Ethical Conflict and the Problem of Consensus,' in *The Concept of Moral Consensus*, K. Bayertz, (ed.), Kluwer Academic Publishers, Dordrecht, Netherlands, pp.75-96.
16. Lieberman, B.A., P.L. Matson, and F. Hamer: 1994, 'The UK Human Fertilisation and Embryology Act 1990–how well is it functioning?' *Human Reproduction*, **9**,9, 1779-1782.
17. Lorber, J.: 1992, 'Choice, Gift or Patriarchal Bargain? Women's Consent to *In Vitro* Fertilization in Male Infertility.' in H. Holmes and L. Purdy, (eds.), *Feminist Perspectives in Medical Ethics*, Indiana University Press, Bloomington.
18. Lyotard, J-F.: 1984, *The Postmodern Condition: A Report on Knowledge*, (trans. G.Bennington and B.Massumi, original text, 1979), University of Minnesota Press, Minneapolis, Mn
19. Moreno, J.D.: 1995, *Deciding Together: Bioethics and Moral Consensus*, Oxford, New York.
20. Morgan, R.L. and D. Lee: 1991, *Blackstone's Guide to the Human Fertilisation and Embryology Act 1990*, Blackstone Press, London.
21. Novaes, S.: 'Beyond Consensus about Principles: Decision-Making by a Genetics Advisory Board in Reproductive Medicine,' in *The Concept of Moral Consensus*, K. Bayertz, (ed.), Kluwer Academic Publishers, Dordrecht, Netherlands, pp. 207-221.

22. Petchesky, R.: 1986. *Abortion and Woman's Choice: The State, Sexuality and Reproductive Freedom*. Verso, London.
23. Price, F.: 1990. 'The Management of Uncertainty in Obstetric Practice: Ultrasonography, In Vitro Fertilisation and Embryo Transfer.' In M. McNeil, I., Varcoe and S. Yearley, (eds.) *The New Reproductive Technologies*, Macmillan, London, pp. 123-153.
24. Progress: Campaign for Research into Human Reproduction: 1989, 'Freedom to Choose: Research into Infertility and Congenital Handicap.' London, U.K.
25. Rose, H.: 1994. *Love, Power and Knowledge*, Indiana University Press, Bloomington.
26. Royal Commission on New Reproductive Technologies: 1993, *Final Report: Proceed with Care*, Canadian Communications Group, Ottawa, Canada.
27. Steinberg, D.L.: 1990. 'The Depersonalization of Women through the Administration of In Vitro Fertilisation.' In M. McNeil, I. Varcoe and S. Yearley, (eds.) *The New Reproductive Technologies*, Macmillan, London, pp. 74-122.
28. *The Times*: April 20, 1990.
29. Wagner, M.: 1989, 'Are In-Vitro Fertilisation and Embryo Transfer of Benefit to All?', *The Lancet* **335**, 1027-1030.
30. Wagner, M. and P. Stephenson: 1993, 'Infertility and In vitro Fertilization: Is the Tail Wagging the Dog?' in *Tough Choices: In Vitro Fertilization and the Reproductive Technologies*, M. Wagner and P. Stephenson, (eds.), Temple University Press, Philadelphia, pp. 1-22.
31. Warden, John: 1990. 'Abortion and Conscience', *BMJ* **301,** 1031.
32. Warnock, Mary: 1983, 'In Vitro Fertilization: The Ethical Issues II,' *Philosophical Quarterly* **33** (132), 238-249.
33. Warnock, M.: 1985. *A Question of Life*. New York, Blackwell.
34. Williams, Bernard, *et al*: 1980, *Report of Committee on Obscenity and Film Censorship* Cmnd. 7772, HMSO, London, U.K.
35. Wolfenden, J.F., *et al*: 1957, *Report of Committee on Homosexual Offences*, Cmnd. 247, HMSO, London, U.K.
36. Yoxen, E.: 1990. 'Conflicting Concerns: The Political Context of Recent Embryo Research Policy in Britain.' In M. McNeil, I. Varcoe and S. Yearley, (eds.), *The New Reproductive Technologies*, Macmillan, London, pp. 173-199.

NOTES ON CONTRIBUTORS

Gebhard Allert, Dr.med., is Scientific Assistant, Department of Psychotherapy and member of the study group for Medical Ethics at the University of Ulm, Germany.

Helmut Baitsch, em. Prof.Dr.med., Dr.rer.nat., is former Director of the Department of Anthropology and Human Genetics, former President of the Universities of Freiburg and Ulm, member of the study group for Medical Ethics at the University of Ulm, Germany.

Robert M. Cook-Deegan, M.D., is Director of the Division of Biobehavioral Sciences and Mental Disorders of the Institute of Medicine, National Academy of Sciences, Washington, D.C., USA.

Anne Donchin, Ph.D., is Professor, Department of Philosophy, Indiana University, Indianapolis, Indiana, USA.

James F. Drane, M.D., is Professor Emeritus, Department of Philosophy, Edinboro University of Pennsylvania, Edinboro, USA.

K.W.M. Fulford, M.D., Ph.D., is Professor of Philosophy and Mental Health, Department of Philosophy, University of Warwick, Coventry, United Kingdom.

J. Chris Hackler, Ph.D., is Professor, Division of Medical Humanities, University of Arkansas for Medical Sciences, Little Rock, Arkansas, USA.

Henk A.M.J. ten Have, M.D., Ph.D., is Professor of Medical Ethics, Department of Ethics, Philosophy and History of Medicine, School of Medical Sciences, Catholic University of Nijmegen, Nijmegen, The Netherlands.

Monika Kautenburger is Academic Councillor, Study-Commission for the Medical Sciences, University of Ulm, Germany.

Akio Sakai, M.D., is Professor of Neuropsychiatry, Department of Neuropsychiatry, School of Medicine, Iwate Medical University, Morioka, Japan.

Hans-Martin Sass, Ph.D., is Professor of Philosophy, Institute of Philosophy, Ruhr-Universität Bochum, Germany, and Senior Research Fellow, Kennedy Institute of Ethics, Georgetown University, Washington, D.C., USA.

Stuart F. Spicker, Ph.D., is Professor of Philosophy and Healthcare Ethics, Massachusetts College of Pharmacy and Allied Health Sciences, Boston; Professor Emeritus, University of Connecticut School of Medicine, USA.

Sandro Spinsanti, Ph.D., is Director, Istituto Giano, Rome, Italy.

Gerlinde Sponholz, Dr.med., Dr.rer.biol.hum., is Scientific Assistant, Department of Medical Genetics and member of the study group for Medical Ethics at the University of Ulm, Germany.

Robert M. Veatch, Ph.D., is Professor of Medical Ethics, The Kennedy Institute of Ethics, Georgetown University, Washington, D.C., USA.

Henrik R. Wulff, M.D., is Chief Physician, Department of Medicine, Herlev University Hospital, and Professor, Department of Medical Philosophy and Clinical Theory, Panum Institute, University of Copenhagen, Copenhagen, Denmark.

Hub A.E. Zwart, Ph.D., is Director of the Centre for Ethics, Department of Philosophy, Catholic University of Nijmegen, Nijmegen, The Netherlands.

INDEX

Philosophy and Medicine

1. H. Tristram Engelhardt, Jr. and S.F. Spicker (eds.): *Evaluation and Explanation in the Biomedical Sciences.* 1975 ISBN 90-277-0553-4
2. S.F. Spicker and H. Tristram Engelhardt, Jr. (eds.): *Philosophical Dimensions of the Neuro-Medical Sciences.* 1976 ISBN 90-277-0672-7
3. S.F. Spicker and H. Tristram Engelhardt, Jr. (eds.): *Philosophical Medical Ethics.* Its Nature and Significance. 1977 ISBN 90-277-0772-3
4. H. Tristram Engelhardt, Jr. and S.F. Spicker (eds.): *Mental Health.* Philosophical Perspectives. 1978 ISBN 90-277-0828-2
5. B.A. Brody and H. Tristram Engelhardt, Jr. (eds.): *Mental Illness.* Law and Public Policy. 1980 ISBN 90-277-1057-0
6. H. Tristram Engelhardt, Jr., S.F. Spicker and B. Towers (eds.): *Clinical Judgment.* A Critical Appraisal. 1979 ISBN 90-277-0952-1
7. S.F. Spicker (ed.): *Organism, Medicine, and Metaphysics.* Essays in Honor of Hans Jonas on His 75th Birthday. 1978 ISBN 90-277-0823-1
8. E.E. Shelp (ed.): *Justice and Health Care.* 1981 ISBN 90-277-1207-7; Pb 90-277-1251-4
9. S.F. Spicker, J.M. Healey, Jr. and H. Tristram Engelhardt, Jr. (eds.): *The Law-Medicine Relation.* A Philosophical Exploration. 1981 ISBN 90-277-1217-4
10. W.B. Bondeson, H. Tristram Engelhardt, Jr., S.F. Spicker and J.M. White, Jr. (eds.): *New Knowledge in the Biomedical Sciences.* Some Moral Implications of Its Acquisition, Possession, and Use. 1982 ISBN 90-277-1319-7
11. E.E. Shelp (ed.): *Beneficence and Health Care.* 1982 ISBN 90-277-1377-4
12. G.J. Agich (ed.): *Responsibility in Health Care.* 1982 ISBN 90-277-1417-7
13. W.B. Bondeson, H. Tristram Engelhardt, Jr., S.F. Spicker and D.H. Winship: *Abortion and the Status of the Fetus.* 2nd printing, 1984 ISBN 90-277-1493-2
14. E.E. Shelp (ed.): *The Clinical Encounter.* The Moral Fabric of the Patient-Physician Relationship. 1983 ISBN 90-277-1593-9
15. L. Kopelman and J.C. Moskop (eds.): *Ethics and Mental Retardation.* 1984 ISBN 90-277-1630-7
16. L. Nordenfelt and B.I.B. Lindahl (eds.): *Health, Disease, and Causal Explanations in Medicine.* 1984 ISBN 90-277-1660-9
17. E.E. Shelp (ed.): *Virtue and Medicine.* Explorations in the Character of Medicine. 1985 ISBN 90-277-1808-3
18. P. Carrick: *Medical Ethics in Antiquity.* Philosophical Perspectives on Abortion and Euthanasia. 1985 ISBN 90-277-1825-3; Pb 90-277-1915-2
19. J.C. Moskop and L. Kopelman (eds.): *Ethics and Critical Care Medicine.* 1985 ISBN 90-277-1820-2
20. E.E. Shelp (ed.): *Theology and Bioethics.* Exploring the Foundations and Frontiers. 1985 ISBN 90-277-1857-1
21. G.J. Agich and C.E. Begley (eds.): *The Price of Health.* 1986 ISBN 90-277-2285-4
22. E.E. Shelp (ed.): *Sexuality and Medicine.* Vol. I: Conceptual Roots. 1987 ISBN 90-277-2290-0; Pb 90-277-2386-9

23. E.E. Shelp (ed.): *Sexuality and Medicine*. Vol. II: Ethical Viewpoints in Transition. 1987 ISBN 1-55608-013-1; Pb 1-55608-016-6
24. R.C. McMillan, H. Tristram Engelhardt, Jr., and S.F. Spicker (eds.): *Euthanasia and the Newborn*. Conflicts Regarding Saving Lives. 1987 ISBN 90-277-2299-4; Pb 1-55608-039-5
25. S.F. Spicker, S.R. Ingman and I.R. Lawson (eds.): *Ethical Dimensions of Geriatric Care*. Value Conflicts for the 21th Century. 1987 ISBN 1-55608-027-1
26. L. Nordenfelt: *On the Nature of Health*. An Action-Theoretic Approach. 2nd, rev. ed. 1995 ISBN 0-7923-3369-1; Pb 0-7923-3470-1
27. S.F. Spicker, W.B. Bondeson and H. Tristram Engelhardt, Jr. (eds.): *The Contraceptive Ethos*. Reproductive Rights and Responsibilities. 1987 ISBN 1-55608-035-2
28. S.F. Spicker, I. Alon, A. de Vries and H. Tristram Engelhardt, Jr. (eds.): *The Use of Human Beings in Research*. With Special Reference to Clinical Trials. 1988 ISBN 1-55608-043-3
29. N.M.P. King, L.R. Churchill and A.W. Cross (eds.): *The Physician as Captain of the Ship*. A Critical Reappraisal. 1988 ISBN 1-55608-044-1
30. H.-M. Sass and R.U. Massey (eds.): *Health Care Systems*. Moral Conflicts in European and American Public Policy. 1988 ISBN 1-55608-045-X
31. R.M. Zaner (ed.): *Death: Beyond Whole-Brain Criteria*. 1988 ISBN 1-55608-053-0
32. B.A. Brody (ed.): *Moral Theory and Moral Judgments in Medical Ethics*. 1988 ISBN 1-55608-060-3
33. L.M. Kopelman and J.C. Moskop (eds.): *Children and Health Care*. Moral and Social Issues. 1989 ISBN 1-55608-078-6
34. E.D. Pellegrino, J.P. Langan and J. Collins Harvey (eds.): *Catholic Perspectives on Medical Morals*. Foundational Issues. 1989 ISBN 1-55608-083-2
35. B.A. Brody (ed.): *Suicide and Euthanasia*. Historical and Contemporary Themes. 1989 ISBN 0-7923-0106-4
36. H.A.M.J. ten Have, G.K. Kimsma and S.F. Spicker (eds.): *The Growth of Medical Knowledge*. 1990 ISBN 0-7923-0736-4
37. I. Löwy (ed.): *The Polish School of Philosophy of Medicine*. From Tytus Chałubiński (1820–1889) to Ludwik Fleck (1896–1961). 1990 ISBN 0-7923-0958-8
38. T.J. Bole III and W.B. Bondeson: *Rights to Health Care*. 1991 ISBN 0-7923-1137-X
39. M.A.G. Cutter and E.E. Shelp (eds.): *Competency*. A Study of Informal Competency Determinations in Primary Care. 1991 ISBN 0-7923-1304-6
40. J.L. Peset and D. Gracia (eds.): *The Ethics of Diagnosis*. 1992 ISBN 0-7923-1544-8

41. K.W. Wildes, S.J., F. Abel, S.J. and J.C. Harvey (eds.): *Birth, Suffering, and Death*. Catholic Perspectives at the Edges of Life. 1992 [CSiB-1]
ISBN 0-7923-1547-2; Pb 0-7923-2545-1
42. S.K. Toombs: *The Meaning of Illness*. A Phenomenological Account of the Different Perspectives of Physician and Patient. 1992
ISBN 0-7923-1570-7; Pb 0-7923-2443-9
43. D. Leder (ed.): *The Body in Medical Thought and Practice*. 1992
ISBN 0-7923-1657-6
44. C. Delkeskamp-Hayes and M.A.G. Cutter (eds.): *Science, Technology, and the Art of Medicine*. European-American Dialogues. 1993 ISBN 0-7923-1869-2
45. R. Baker, D. Porter and R. Porter (eds.): *The Codification of Medical Morality*. Historical and Philosophical Studies of the Formalization of Western Medical Morality in the 18th and 19th Centuries, Volume One: Medical Ethics and Etiquette in the 18th Century. 1993 ISBN 0-7923-1921-4
46. K. Bayertz (ed.): *The Concept of Moral Consensus*. The Case of Technological Interventions in Human Reproduction. 1994 ISBN 0-7923-2615-6
47. L. Nordenfelt (ed.): *Concepts and Measurement of Quality of Life in Health Care*. 1994 [ESiP-1] ISBN 0-7923-2824-8
48. R. Baker and M.A. Strosberg (eds.) with the assistance of J. Bynum: *Legislating Medical Ethics*. A Study of the New York State Do-Not-Resuscitate Law. 1995 ISBN 0-7923-2995-3
49. R. Baker (ed.): *The Codification of Medical Morality*. Historical and Philosophical Studies of the Formalization of Western Morality in the 18th and 19th Centuries, Volume Two: Anglo-American Medical Ethics and Medical Jurisprudence in the 19th Century. 1995 ISBN 0-7923-3528-7; Pb 0-7923-3529-5
50. R.A. Carson and C.R. Burns (eds.): *Philosophy of Medicine and Bioethics*. A Twenty-Year Retrospective and Critical Appraisal. 1997
ISBN 0-7923-3545-7
51. K.W. Wildes, S.J. (ed.): *Critical Choices and Critical Care*. Catholic Perspectives on Allocating Resources in Intensive Care Medicine. 1995 [CSiB-2]
ISBN 0-7923-3382-9
52. K. Bayertz (ed.): *Sanctity of Life and Human Dignity*. 1996
ISBN 0-7923-3739-5
53. Kevin Wm. Wildes, S.J. (ed.): *Infertility: A Crossroad of Faith, Medicine, and Technology*. 1996 ISBN 0-7923-4061-2
54. Kazumasa Hoshino (ed.): *Japanese and Western Bioethics*. Studies in Moral Diversity. 1996 ISBN 0-7923-4112-0
55. E. Agius and S. Busuttil (eds.): *Germ-Line Intervention and our Responsibilities to Future Generations*. 1998 ISBN 0-7923-4828-1
56. L.B. McCullough: *John Gregory and the Invention of Professional Medical Ethics and the Professional Medical Ethics and the Profession of Medicine*. 1998 ISBN 0-7923-4917-2
57. L.B. McCullough: *John Gregory's Writing on Medical Ethics and Philosophy of Medicine*. 1998 [CiME-1] ISBN 0-7923-5000-6

Philosophy and Medicine

58. H.A.M.J. ten Have and H.-M. Sass (eds.): *Consensus Formation in Healthcare Ethics*. 1998 [ESiP-2] ISBN 0-7923-4944-X

KLUWER ACADEMIC PUBLISHERS – DORDRECHT / BOSTON / LONDON

MIX
Papier aus verantwortungsvollen Quellen
Paper from responsible sources
FSC® C105338
FSC
www.fsc.org

If you have any concerns about our products,
you can contact us on
ProductSafety@springernature.com

In case Publisher is established outside the EU,
the EU authorized representative is:
Springer Nature Customer Service Center GmbH
Europaplatz 3, 69115 Heidelberg, Germany

Printed by Libri Plureos GmbH
in Hamburg, Germany